An introduction to
the history of
virology

A selection of tulip flowers showing various distinctive patterns characteristic of infection with tulip mosaic virus. The original colour plate, paint-
ed by Philip Reinagle, R.A., is included in *The Temple of Flora* (London,
1799), by John Robert Thornton, a graduate of Trinity College, Cambridge, and Guy's Hospital Medical School, who became a lecturer on
medical botany at the United Hospitals of Guy and St Thomas'.
Thornton's consuming interest in botanical engaving led him into over-
ambitious publishing ventures (including *The Temple of Flora*), and his
later life, until his death in 1837, was beset with financial difficulties.

An introduction to the history of virology

A. P. WATERSON

Professor of Virology
Royal Postgraduate Medical School
University of London

&

LISE WILKINSON

Research Fellow
Department of Virology
Royal Postgraduate Medical School
University of London

'Many shall run to and fro, and
knowledge shall be increased.'
Daniel xii.4

CAMBRIDGE UNIVERSITY PRESS

Cambridge
London   New York   Melbourne

Published by the Syndics of the Cambridge University Press
The Pitt Building, Trumpington Street, Cambridge CB2 1RP
Bentley House, 200 Euston Road, London NW1 2DB
32 East 57th Street, New York, NY 10022, USA
296 Beaconsfield Parade, Middle Park, Melbourne 3206, Australia

© Cambridge University Press 1978

First published 1978

Printed in Great Britain
at the University Press, Cambridge

Library of Congress Cataloguing in Publication Data
Waterson, A. P.
An introduction to the history of virology.
Bibliography: p.
Includes indexes.
1. Virology – History. I. Wilkinson, Lise,
joint author. II. Title.
QR359.W37   576'.64   77-17892
ISBN 0 521 21917 5

Contents

'Of the three adjectives which might be employed to express the idea that contagium consists of particles, viz. molecular, corpuscular and particulate, the former, although more familiar, are objectionable, because the words from which they are derived have already been appropriated. "Molecule" is understood in physics to mean a hypothetical body of indefinite minuteness, while "corpuscle" is used in anatomy to denote bodies of much larger size than those in question.'

Burdon Sanderson, 1869

'In order to determine whether the virus [tobacco mosaic virus] should be considered as corpuscular or as dissolved the following experiment was devised... Hence there appears to be little doubt that the contagium must be regarded as liquid, or perhaps better expressed, as water-soluble.'

M. W. Beijerinck, 1898

'...this suggestion also had its repercussions, for whether or not viruses were particulate was solemnly debated, without anyone apparently asking what they could be if they were not particles of some size or other.'

F. C. Bawden, 1964

Aims, possibilities and objectives

by A. P. Waterson

The word virology itself is very recent, and the science of virology not much older, so that some may question even whether it has much of a history to write. In fact, as a study in the history of science it has proved rewarding. Its relatively recent origin means that the sources are copious and, on the whole, accessible, even if they are somewhat diffusely spread. Many of the people involved in some of the quite early work are still alive. Very much has happened in a short time, so that the rate and acceleration of growth of knowledge have been so fast that the direction and flow of thought have been easy to follow. The origins are various, but easy to discern. Botany, plant pathology, human and veterinary medicine, and especially that activity known as hygiene on the continent of Europe and as bacteriology in Great Britain and the United States, whether by doctors or veterinarians, have all contributed. As time went on, and momentum was gathering, genetics, protein chemistry, cytology and molecular biology all became involved.

It was in fact a consideration of virology as an offshoot of bacteriology, and an experience of the difficulties of working in a young subject under the terms of reference of an old, which supplied much of the stimulus to initiate the current study. Indeed, the development of virology could, with profit, be studied by historians as a model of the emergence and establishment of a new branch of biological science. The shifts of academic and organizational thrust, the ebb and flow of emphasis, and the forces behind them, comprise in themselves a valuable model system. The rise and fall of research institutes, the inception of new academic courses, the effects of rigidly determined medical curricula, the interplay between pure and applied science, have all played a part in the genesis of virology. Not least, the subject could serve as a *locus classicus* for anyone wishing to study the relation between scientific research and medical practice, using the term in its widest sense. Such a study would necessarily involve a consideration of the cross-linkages between disciplines and groups of disciplines, especially between the physical and chemical, on the one hand, and the biological, on the other. For example, the Max Planck Institute for Virus Research at Tübingen sprang not from a microbiology laboratory but from an institute for research in biochemistry.

Surprisingly, one of the most potent, if somewhat subterranean, influences has been that of the atomic physicists of the 1920s, principally Erwin Schrödinger, whose book *What is life?* proved seminal to modern molecular biology and to viral genetics. Ostensibly it was because they hoped to find 'new laws' that such investigators as Schrödinger and Delbrück were interested in biological phenomena. The preoccupation of physicists at this time with sub-atomic particles may also have given them a fellow feeling with microbiologists dealing with what seemed, on their scale, to be infinitely, and indeed impossibly, small particles.

This very element of near impossibility with which the early virologists were faced makes the subject a fertile one for the historian on two accounts. The principal difficulties were those intrinsic to the study of very small biological particles, and the pioneers had little at hand in their armamentarium. However, sooner or later, techniques did emerge. Centrifuges grew in speed and power, and microscopes were developed until their resolution and magnification went beyond what the specimen and the preparative techniques merited. In the meantime, the methods available were stretched to their limits, and it would be a rewarding study to find out how far the problem stimulated the technical development, and vice versa. Certainly in one field, that of cell and tissue culture, the demands of virology supplied a powerful *vis a tergo* to the technique, which at one time was in danger of becoming an inward-looking mystique.

The second reason why the early experimental inaccessibility of the small particles is of interest is that it is now possible, by hindsight, to judge how correct, or otherwise, were the ideas and the concepts formed by the early virologists about the agents they studied. In other words, their mental concepts of what they were studying can be examined, and the development of these concepts studied, together with the influences which played upon them, and the reasons for which they elaborated the pictures they did of what they could not see. The evaluation of size from indirect evidence and the beneficial effect of an increasingly quantitative approach plays an interesting part in the development of thought in this area. What is of particular interest is how early this began to happen. Beijerinck was thinking in molecular terms before the turn of the century, but was not matched by the organic chemists of his day. Ideas on the size of large molecules came to the fore surprisingly early, but the birth pangs of the macromolecular concept make surprising reading today in an age of polymer chemistry, and are an object lesson in the difficulty of accepting a scientific concept when it does not fit in with what has been accepted, entrenched and even cherished.

In the light of all this, it is clear that several different histories of virology might have been written. Of course, each would embrace the same general series of events but each would concentrate on one or other facet of the whole story. For example, it could have taken the form of an account of the various academic and other environments in which virology grew up. This would be of great interest to anyone interested in the founding of new subjects and the organizational obstacles which have to be overcome when this happens. It would be a study of how a subject becomes a subject, and what justification it may have for enjoying all that this entails in terms of accommodation, research funds and, within a university, teaching facilities. Again, the history might have concentrated on the experimental basis alone, i.e. on the discoveries made, the technical problems involved and the technical solutions to them, including the influence of such techniques on fields outside virology. Another approach would have been the biographical one. It would have concerned itself with the personalities involved, with what made them incline to this particular field, what made them excel in it, and, when they failed and were held up, why this should have happened. Of course, all these three elements are essential as ingredients, but the proportions used, and their quantities relevant to each other, would vary from one account to another.

In fact, the approach chosen has been different from all of these, although in varying degrees it has embraced each of them. The underlying theme of this book, which is to be viewed as an *aditus* to the subject as a whole, is the evolution of the present concept of a virus. This present-day concept is essentially a chemical one, but that is not to imply that it is divorced from the mainstream of *biological* thinking, but, because it is chemical, and because definition cannot proceed further than molecules, it is to be regarded as definitive, and the present time as a good vantage point from which to look back, so that we can better look around and better look forward. It is therefore essentially apocalyptic, in the proper sense of the word, i.e. revelatory. It is possible to study the virological scientists of the past, and to know in the light of present knowledge exactly what they were studying, even though this was not revealed to them in their day. It is, in other words, the story of the progressive unveiling of the nature of the virus particle.

This history is therefore essentially, and indeed deliberately, *conceptual*. It is a study of ideas and concepts, and the inter-reactions between these, on the one hand, and experiment and technique, on the other. That it is inextricably interwoven with the administrative, the technical and the personal goes without saying, and to compose this history involved in itself the administrative, the technical and the personal. It involved the administrative, because it was initiated in the belief that the work could best be done in an orthodox academic department of virology, but by a full-time and committed historian, and this is how it came into being. It involved the technical, because it was necessary to devise a historical technique by which the necessary examination, analysis and synthesis could be achieved. It is apposite here to outline this method. Four well-studied viruses were chosen. These were fowl plague (avian influenza A), tobacco mosaic, rabies and smallpox.[1] All were early on the scene, all have been exhaustively studied. Their history was

[1] These and an introductory paper have been published in *Medical History*.

studied with particular attention to the ideas in the minds of the succession of scientists concerned about the nature, or essence, of the agent with which they were dealing. What was it in terms of known biological phenomena? What was its chemical constitution? How big was it? Above all, was it something *sui generis,* or was it another example of something already known? This provided four bodies of information, each in chronological sequence, so that the conceptual activities of virologists on at least four agents could be compared at any one point in time, especially in the last 90 years. This provided a frame of reference of fixed points in time and thought, against which other material could be assessed. This was then collated with the general discourses and reviews on the whole field which have appeared from time to time, and such 'battles long ago' as the now burnt-out controversy as to whether viruses are 'living' or 'non-living', seen in perspective. The resulting synthesis constitutes the main corpus of this book.

The third element has been the personal. A catalogue may be compiled by a computer, but stories can be told only by a person. Inevitably, as persons too, individual virologists figure in these pages, but it is their thinking as much as their doing with which we have been concerned, because the lesser (doing) is included in the greater (thinking). Biographical details have been relegated to an appendix, not because they are unimportant, but to prevent them from obstructing the main flow of the narrative. Narrative is perhaps a key word, because this is, above all, a story. It is to be hoped that, as a story, it will be read with enjoyment no less than, as a history, it will be studied with interest.

Acknowledgments

The present book grew out of research on the development of the virus concept, supported by a grant from the Wellcome Trust, which has been described in a series of papers published in *Medical History*. The authors wish to thank Dr Edwin Clarke, Director of the Wellcome Institute for Medical History and editor of *Medical History*, for permission to include some of this material, especially in Chapters 4 and 10, and also for much friendly advice and discussion. We are also grateful to all the friends and colleagues at home and abroad who have patiently answered questions in connexion with the present work; to the librarians of the Wellcome Institute Library and the Royal Society of Medicine, and of the Royal Society for help with specific points in the biographical notes of Fellows; to Mrs Lois King for preparation of the typescript for the press; to Mr Peter Lister for preparation of photographs for some of the illustrations; and to the journals and institutions quoted for permission to reproduce individual illustrations.

Inclusion of the colour frontispiece was made possible by a grant from the Wellcome Trust which is gratefully acknowledged. The plate was reproduced from the original volume of *The Temple of Flora* in the Library of Eton College, by courtesy of its Provost and Fellows.

1. Early terminology and underlying ideas

The term *virology* has become part of our vocabulary only gradually during recent decades. The first edition of S. E. Luria's *General Virology* was published in 1953, and the first issue of the journal *Virology* appeared in 1955. Viruses, on the other hand, have been with man for a very long time, and so has the word *virus*, although its connotations and usage have varied over the centuries.

Even the most liberal interpretation of accuracy of definition would allow us to date the history of any true concept of viruses only from the end of the nineteenth century. If we include epidemiological aspects in our historical considerations, we are still on uncertain ground with most virus diseases for the better part of our calendar, with one outstanding exception. Rabies, whose ravages in terms of human lives lost or affected are slight compared with the major scourges of mankind, has been meticulously described and recorded for more than two millennia. Its mysterious aetiology, its ability to transform a friendly, domesticated dog into a raging, vicious beast, its long and ill-defined incubation period, and the all too well-defined and infinitely distressing symptoms preceding the inevitable fatality of the frank clinical disease in man – all combined to present a picture which by its very terror made it irresistible to the writers and thinkers of antiquity. Comparisons of early Roman and Greek descriptions with recent case reports suggest that rabies virus has changed little if at all during the intervening years. This distinguishes it from most other pathogenic viruses known today, and gives it a unique position in history. Thoughts on the aetiology of rabies have been proffered to the world for a very long time, and the disease itself can be traced back at least

I

as far as Aristotle,[1] who informed his readers that '...rabies drives the animal mad, and any animal whatever, excepting man, will take the disease if bitten by a mad dog so afflicted; the disease is fatal to the dog itself and to any animal it may bite, man excepted' (17). Later commentators, many of whom were reluctant to question the credibility of the great philosopher, have been puzzled by the apparent reference to the insusceptibility of man, and have offered varying explanations,

CANIS RABIDVS.

Fig. 1. Sixteenth-century woodcut showing men attempting to overpower a rabid dog. From Dioscorides (1566), *Acerca de la materia medicinal*, Salamanca: M. Gast. (Courtesy of the Trustees of the Wellcome Foundation.)

ranging from an early change in the syndrome to the more likely one, by Fracastoro (159) that Aristotle merely meant to draw attention to the fact that not all those bitten by a mad dog would necessarily develop the clinical disease.

A more detailed and remarkably accurate description of the

[1] Some other claims would appear to be less well founded. That Democritus 'considered the disease an inflammation of the nerves' is probably explained by an over-optimistic translation of Caelius Aurelianus' Latin rendering of Soranus' Greek texts (19). Democritus' contemporary, Euripides, is said to have been subjected to salt-water therapy by being thrown into the sea following a dog bite. This type of treatment was long adhered to – *faute de mieux* – and Boissier de Sauvages attempted to explain it in chemical terms in the eighteenth century (see 443; 444).

rabies syndrome was provided by Celsus, who lived and wrote during the heyday of the Roman Empire in the first century A.D. One particular sentence from Celsus' chapter on rabies has been extensively quoted. 'Especially if the dog was rabid, the virus must be drawn out with a cupping glass' (73). Of course no one would suggest that Celsus with these words identified the pathogen of rabies as a virus in modern terms; nevertheless, the apparent distinction he draws between the use of 'virus' to denote the agent of rabies and, later in the text, of 'venenum' to describe the poisonous principle from snakes may have been not wholly accidental. It may reflect the alternative meaning of 'slimy liquid' for 'virus' in Latin, and hence Celsus' awareness that the agent of rabies was transmitted through the (slimy) saliva of the rabid dog.

For centuries after Celsus, the term 'virus' was used casually as a synonym for poison or venom, until with the growing awareness of transmission of disease in the eighteenth and nineteenth centuries it eventually acquired the meaning of an infectious agent. The gradual acceptance of this usage in medical literature ran parallel to the development of the twin concepts of infection and contagion, and both owed much to another virus disease of exceptional historical interest. Unlike rabies, smallpox through the centuries has had a marked influence on the course of social and political history.

In terms of devastation frequently wreaked in mediaeval communities, smallpox is equalled among the infectious diseases only by bubonic plague, and this is one of the obvious reasons why smallpox played a major role for the popular acceptance of the principle of disease transmissibility. At the same time, the early history of smallpox presents a far greater problem to the historian than does rabies; the difficulties inherent in any attempt to establish differential diagnosis of smallpox and other eruptive fevers, such as measles, chickenpox, German measles, and even scarlet fever, on the basis of often vague and incomplete descriptions made several hundred years ago in Arabic or classical or mediaeval Latin, are daunting and frequently insurmountable. Even Fracastoro, whose perceptive sixteenth-century comments on contagion command the respect of anyone interested in the early development of this concept, had

3

A
DECLARATION
OF SVCH GREIVOVS
accidents as commonly follow
the biting of mad Dogges,
together with the cure
thereof,

BY
THOMAS SPACKMAN
Doctor *of* Phyſick.

LONDON
Printed for *Iohn Bill* 1613.

Fig. 2. Self-explanatory title page of Thomas Spackman's (1613) treatise on rabies, London. (Courtesy of the Royal Society of Medicine.)

4

little to contribute in his chapter on smallpox and measles. Although he fitted what he called collectively 'poxes and measles' (variolae and morbilli)[1] into his general theory of contagion with the remark '...what exhales from the putrefaction...is a germ of contagion for another individual...' (159) he was at pains also to associate himself with the idea inherited from the Arabic writers, notably Rhazes, that the eruptive fevers were a salutary and necessary form of purgation suffered to greatest advantage in childhood, the earlier the better.

But the lessons of the major smallpox epidemics of the seventeenth and eighteenth centuries, when particularly virulent strains struck the European continent, were obvious. Something was transmitted from person to person, from house to house. The practice of variolation may have had its origins in pagan, superstitious attempts to transfer evil afflictions to a third party. It was introduced into Europe in its established form from the Middle East, and, from 1721 onwards, was made fashionable in England and on the European continent when it was championed by Lady Mary Wortley Montagu[2] after her return from her husband's Constantinople Embassy. As a lesser evil, variolation enjoyed considerable popularity in informed circles until Jenner presented his safer alternative of vaccination in 1798.

The problems of the severe epidemics and the controversy surrounding the variolation procedures formed the basis for much medical literature in the late seventeenth and throughout the eighteenth century. In the course of their considerations, some authors allowed themselves space to speculate on the nature of the contagion; during the eighteenth century there was a growing tendency to label the principle transmitted 'virus'. Thomas Fuller, in 1730, included quite advanced views in his account of eruptive fevers, and wrote: 'The chief and commonest Way of taking the contagious Fevers, Small-Pox and Measles, is by Infection; that is, by receiving with the Breath,

---

[1] Fracastoro's chapter on poxes and measles is not among the most lucid in *De contagione et contagionis morbis et eorum curatione*, and his distinction between variolae and morbilli is vague (159).

[2] An interesting biographical sketch of Lady Mary has been supplied by Robert Halsband who has unearthed a letter suggesting that Timonis, who with Pylarino first introduced variolation in Europe and to the Royal Society (448; 361), was physician to the Embassy household at the time Lady Mary conceived the idea of variolation for her own children (196).

5

or thro' the Pores, such virose Corpuscles, as are peculiar for the Breeding of them' (167). About this time the term even found its way into more unexpected areas of pathology, as when Lorenz Heister, describing mastectomy in *Institutiones chirurgicae* in 1739, could be seen to anticipate modern cancer studies in his discussion of the precautions which should be taken with '...the blood which is infected by the cancer virus' (204).

The great but sometimes neglected forerunner of Jenner, Angelo Gatti, published his *Réfléxions* on variolation in 1764. His observations on the nature of the infection and on the exigence of some means of attenuating 'the variolous virus' (174, 175), have a prophetic quality which leaves the reader regretting that he died just too soon to learn of Jenner's discovery. Gatti died in January 1798, and never knew that Jenner had hit upon the 'cow-pox virus' to fulfil the need for an attenuated form of the 'variolous virus' which Gatti had been hoping to achieve for more than 30 years.

The impact of Jenner's classic paper was felt outside medical literature as well. In 1806, Thomas Jefferson wrote to congratulate Jenner on his great achievement, adding: 'Yours is the comfortable reflection that mankind can never forget that you have lived. Future nations will know by history only that the loathsome smallpox-pox has existed and by you has been extirpated.'[1] That this was to take nearly 200 years would have seemed an excessively pessimistic estimate to both Jefferson and Jenner. Public awareness of the method was such that the terminology of the vaccinator even found its way into the metaphors of the Lake poets. Southey wrote in his introduction to the 'Remains' of H. K. White in 1807: 'As if there were not enough of the leaven of disquietude in our natures, without inoculating it with this dilutement – this vaccine virus of envy.'

By the early nineteenth century the term 'virus' was certainly well established in medical writing and used to describe a wide variety of infectious agents. The usage persisted throughout the rise of bacteriology during the nineteenth century, with an interesting diversion in mid-century, reflecting the uncertainties and confusion accompanying the satisfaction of identifying

[1] Jefferson's letter is quoted in full by Jenner's faithful friend Baron in the biographical study published in 1838 (26).

the causative organisms of some communicable diseases, while others, equally obviously transmissible, defied identification. If it has no longer more than curiosity value, it involves the names of a cluster of men who in their different ways left an indelible mark on the development of medical methodology and the considerable results it produced.

Claude Bernard erected a framework for experimental medicine supported by a philosophical tradition with its roots in Descartes and, above all, in Pascal.[1] It enabled him to put a crack in the defences of Virchow's early anticontagionism. He cited the results obtained by pure observation in the case of the skin mite *Acarus* as justification for a more extensive use of experimentation in pathology. In his arguments he stressed the advantages of observations made on the *Acarus*, visible to the naked eye, to theories presuming the existence of '...a virus, a creature of reason'. At the same time, Davaine in the 1860s was using a mixture of observation and experimentation to track down what he referred to variously as the virus, the toxic agent and, finally, the bacteridium of anthrax. In the absence of suitable man-made filters, Davaine demonstrated the difference between infected and non-infected blood by the ingenious use of that most efficient natural membrane filter, the guinea pig placenta (99). Koch's discovery of the spore formation by *Bacillus anthracis* helped to explain some of the inconsistencies in Davaine's data (246). Chauveau, working at the same time on vaccinia, could boast no definitive results. He resorted to a theory which, although to some extent anticipating the discovery of bacterial toxins, helped to confuse further the already obscure use of the terminology in the 1860s. Chauveau distinguished between virulent diseases, caused by viruses, and contagious diseases, in which a known and recognized parasite was transmitted; the virus remained a mysterious entity, able to exert its pathogenic influence only through association in solution with certain particulate but unspecified elements. In 1868, Chauveau identified these 'elementary bodies' (*granulations élémentaires*) as the seat of the pathogenic activity, thus introducing the term which was far to outlast Chauveau's work and

[1] Bernard's contribution and philosophy and their background have been discussed in papers and books by a number of authors (see, for example, 400 and 477).

survive until such a time when it could be applied to a more realistic unit.[1]

In 1868, Keber in Danzig, searching for micro-organisms – he called them 'bodies of inflammation' (*Eiterkörperchen*) – in vaccinia lymph, but without recourse to staining, saw 'molecules of hardly measurable tenuity' (241). He filtered the lymph through 'Swedish filter paper', expecting the 'inflammation bodies' to be retained and the filtrate to be non-infectious. To his astonishment, he found that the filtrate produced typical vaccine vesicles upon inoculation. Subjecting the filtered lymph to microscopic examination, he then noticed that although the cellular impurities had been removed, there remained 'a number of nuclei and molecules'. What he called molecules may well have been what Buist stained and called spores of micrococci (56); if they were indeed virus particles, Keber, who undoubtedly used the word in its early sense of 'small particles', could not have imagined how apt the term would one day prove to be.

The years 1865–66 had seen the culmination of the disastrous outbreak of cattle plague, or rinderpest, in Britain. The wise warnings of John Gamgee, who had implored the authorities to impose quarantine and slaughter regulations both before and during the spread of the disease in the British Isles, were disregarded, at an appalling cost to the nation. Gamgee later left the country, a deeply discouraged man. But before he left, he wrote a volume on the cattle plague which in addition to his many practical observations contains far-sighted theoretical reflections. On the nature of cattle plague Gamgee wrote:

Like most animal poisons, the Rinderpest virus is reproduced with marvellous rapidity in, and discharged from, the bodies of sick animals. The breath of a sick ox inspired by a healthy animal, and the solid products of the disease, seem to be alike capable of inducing the malady; and antidotes are applied too late when an attempt is made to reach the poison in the animal's system. I know of no antidote to be used internally...I have no faith in our ever reaching the virus with effect in the living animal... (173)

Words worth remembering when we reflect on how little progress has been made since in the field of anti-viral chemo-

[1] Chauveau proved his point by an ingeniously simple diffusion experiment (78; Fig. 3). The same principle had been employed by Burdon Sanderson in experiments with cattle plague (57).

found that the subsidence, which consisted altogether of leucocytes, was entirely inactive when employed for vaccination. The question therefore lay between the granulations and the serum, and could only be determined by separating the one from the other, and testing each apart. It had already been found that this could neither be done by filtration nor by subsidence. M. Chauveau therefore had recourse to diffusion ;* but the ordinary method of diffusion was not applicable. For the quantity of vaccine liquid which can be obtained for experiment is so exceedingly small that it could not be introduced into a diffusion cell. To obviate this difficulty M. Chauveau operates as follows : he introduces the vaccine from the capillary tube in which it is collected from the arm, into an eprouvette of the form and size shown in the margin, great care being taken to

APPENDIX.

No. 11.
On the Intimate Pathology of Contagion, by Dr. Sanderson.

His method of separating the particles of vaccine from the soluble constituents by diffusion.

FIG. 2.

M. Chauveau's Eprou-
vette for diffusion.
a. Layer of vaccine.
b. Supernatant water.

prevent the liquid from touching the sides of the glass in the act of filling. A drop of water is added, with similar precautions, and the whole allowed to stand for 24 hours. Although the two liquids are thus brought into contact without the intervention of a membrane, yet if the manipulation is dexterously performed they do not mix with each other, except-ing in the immediate neighbourhood of the surface of junction. All the soluble constituents of the vaccine, however, pass upwards into the water. At the end of 24 hours the most superficial layer is removed by dipping in it the end of a fine capillary tube. The liquid in the tube is submitted to microscopical examination, and at once tested for albumen by heat and nitric acid. If it is found, as is invariably the case, to be charged with albumen, it is concluded that the diffusion is complete, for it is clear that if that constituent which possesses least diffusibility of any, has found its way into the most superficial layer, all the others must have done so likewise. The use of the microscopical examination is to show that the liquid is absolutely free from particles. The next step is to compare the two strata of liquid with reference to their activity. For this purpose children or heifers are vaccinated on one side of the body with liquid from the most superficial stratum, collected in the same way as the first specimen was taken for testing ; while a similar number of punctures are made in the other side with undiluted lymph of the same kind as that employed for diffusion. The appearance in due time of a normal vesicle of cow-pox for every punc-ture made with the undiluted lymph, combined with the complete failure of the punctures made with the diffused liquid, shows that the latter, although it contains all the soluble constituents of vaccine in sufficient abundance, is wanting in those on which its activity depends.

The importance of these results appeared to render it necessary to repeat the experiments. As, however, the method did not seem to be perfect, I directed my attention first to the object of improving it. Thus, I thought it particularly desirable that the relative proportions of water and vaccine employed should be precisely ascertained in each

---

* M. Chauveau was, perhaps, not the first to apply the method of diffusion to the analysis of contagious liquids. I had myself made experiments of the same nature in my investigation of cattle plague. The end I had in view in these experiments was to determine whether the virulent principle was a colloid or crystalline substance. With this view I diffused liquids known to be infecting, derived from plague-stricken animals, by the ordinary method, *i.e.*, through parchment paper, and found that while the diffusate was wholly innocuous, the liquid that remained behind retained its activity. I inferred that all contagia were probably colloid substances. Diffusion was em-ployed by M. Chauveau and myself for the same purpose ; but the ideas as to the nature of contagium which led him to adopt it were much more complete than mine.

Fig. 3. Chauveau's diffusion experiment (note 1, p. 8) in an illus-tration by Burdon Sanderson, with the author's footnote explain-ing his own use of the principle in experiments with cattle plague (57). (Reproduced by courtesy of the Royal Society of Medicine.)

therapy, and that, vaccination apart, a policy of isolation and slaughter remains the preferred weapon against most cattle diseases.

Complete understanding of the aetiology of anthrax came with the introduction of a methodological tool which became invaluable in bacteriological studies, and which before the end of the century was directly responsible for the emergence of the concept of the virus as a filterable entity. It seems that unglazed white clay cells connected to an air pump were first used in bacteriological work in 1870–71 by Edwin Klebs (see 380), the instigator of so many nineteenth-century discoveries. Six years later, Pasteur used a plaster of Paris filter in a similar way. With it, and with the collaboration of Joubert, Pasteur moved into the field of medical bacteriology, into anthrax studies, and all that was to follow. Pasteur and Joubert removed the last doubts surrounding the aetiology of anthrax (353). They concluded that there was no need for Chauveau's distinction between virulent and contagious diseases; and after adopting the word 'microbe', coined by Sédillot in 1878 (407), Pasteur could declare '*Tout virus est un microbe*' in 1890 (348).

From then on, through the pioneering work on vaccines against anthrax, chicken cholera and finally rabies, Pasteur, and anyone else engaged in studies in pathology, used 'virus' to denote any infectious agent, bacterial or otherwise, which produced immunity upon recovery. Although Pasteur searched long and in vain for the pathogen of rabies, there is no evidence to suggest that he suspected it of being in any way essentially different from the pathogenic microbes he had characterized over the years, except perhaps with regard to size. His attitude to the term and the concept is summed up in the short sentence quoted above: every virus is a microbe.

In 1897, two close associates and former pupils of Robert Koch headed an inquiry into foot-and-mouth disease, a perennial threat to the German cattle industry (278). The discovery that the infective principle passed easily through the pores of filter candles took them by surprise. Yet the notion of an infectious agent differing in its essential nature from the organisms so ably described and classified by their mentor came no more easily to the staff of the Institut für Infektionskrankheiten in Berlin than to the members of the Institut Pasteur in Paris. Loeffler and Frosch, painstakingly investigating foot-and-mouth disease, inevitably drew the conclusion that its causal agent was a microbe of extreme tenuity.

It is not surprising, given all the facts, that the medical bacteriologists at this time were unlikely to see beyond the microbial concept of pathogens. It is on the other hand very remarkable that a botanical microbiologist, within a year of the publication of the official report on foot-and-mouth disease, should have tentatively suggested an interpretation, which has ultimately proved correct, of his unexpected results with a virus disease of the tobacco plant. Until this time, the concept of infectious plant disease had received scant attention compared to its counterparts in human and animal disease. Fracastoro, justly famed for ideas on contagion which were very much ahead of his time, was aware that certain types of contagion would attack plants and fruits, and that they were specific to plants and not transmitted to domestic animals (159). But he was, after all, a physician, and not concerned with specific plant diseases. In order to find early records of a plant virus disease, it is necessary to look outside the literature of pathology, and to consider the evidence presented by plant breeders and art historians. Art historians and historians of medicine occasionally come together on the subject of human afflictions as depicted in works of art through the centuries. Specimens from the vegetal world have usually found a place in works of art only when intact and healthy. There is one exception. The variegated or 'broken' tulip, popular with the masters of European flower painting from within a few years of its introduction to European gardening[1] little more than 10 years after the publication of Fracastoro's treatise on contagion, owes its characteristic colour patterns to a virus disease. The attractions of its often spectacular manifestations in flowers of affected plants have outweighed the apprehension of gardeners even after the infectious origin of the condition has become known; and the

[1] Tulips were probably first brought to Europe by Ghislain de Busbecq, Austrian ambassador to the court of the Ottoman sultan Soliman the Magnificent, around 1555. Busbecq passed on seeds and bulbs to the great Dutch botanist Carolus Clusius, whose descriptions and illustrations (Fig. 4) suggest that tulip mosaic virus was introduced into Europe with the original consignment of tulip bulbs (see 474).

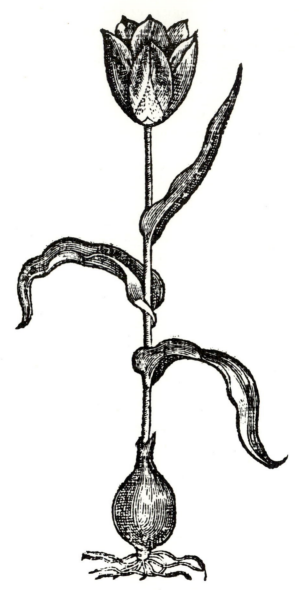

Fig. 4. Variegated tulip, woodcut in C. Clusius (1583), *Rariorum aliquot stirpium, per Pannoniam, Austriam & vicinas quasdam provincias observatarum historia,* Antwerp: Plantin (note 1, p. 11) (Reproduced by courtesy of the Wellcome Trustees.)

diseased varieties, euphemistically christened 'Rembrandt' tulips, still enjoy much popularity with bulb growers.

Throughout the rise of medical and veterinary bacteriology in the nineteenth century, plant pathologists were largely pre-occupied with the study of diseases of fungal aetiology. The notion that plants were susceptible to bacterial attack was only reluctantly accepted towards the end of the century, primarily due to the work of American plant pathologists. For the better part of the nineteenth century, the majority of those concerned believed that causal agents of infectious plant disease were almost exclusively fungal in origin. Some thought that the acidity of their cell contents and the cellulose of their walls rendered plants impervious to bacterial attack, a view promoted in a German textbook as late as 1882. But by then, other opinions were coming to the fore.

The results which were to show the way came from work on a disease of the tobacco plant. It had been under scrutiny for 20 years in Holland and in Russia, and had been named tobacco mosaic disease after the characteristic pattern produced in the leaves of affected plants, when M. W. Beijerinck published his own experimental data and offered an interpretation which was to have ramifications far beyond the world of plant disease (36). D. I. Ivanovski had shown as early as 1892 that the pathogen was small enough to pass through the pores of filter candles impervious to known bacteria (227). Loeffler and Frosch showed the same to be true of the agent of foot-and-mouth disease. They all assumed that they were dealing with microbes of extreme tenuity, different from known micro-organisms only in size. Beijerinck, interpreting his own extensive experiments, trans-cended all the known facts of microbiology and created a con-cept of the filterable virus so bold and precocious that it could not be confirmed in his lifetime.

2. The methodological background

The concept of the filterable virus was formed in the last decade of the nineteenth century. In a period rich in inventions and discoveries, its *fons et origo* was the bacteriological filter developed as an essential tool for the new science of bacteriology. In his trenchant studies on cholera, John Snow had observed (421; 422) that some types of soil deposits were more effective than others in reducing the amounts of infective material from drinking water passing through them. Snow remarked on the relative merits in this respect of gravel, limestone, clay, etc.

Also in the early 1850s, Thomas Graham carried out his experiments on diffusion and osmosis which earned him a reputation as a founder of colloid chemistry (see 88). To demonstrate the validity of his law for the diffusion of gases, he used 'Graham's tube', a simple glass tube plugged at one end with plaster of Paris; moving on to osmosis, he fitted the glass tube, with a cover of guttapercha, to the mouth of a porous clay cylinder. Shortly afterwards, the deliberate use of clay as a filtering material was taken up by the pathologists as a corollary of the intense interest in the investigation of anthrax. The anthrax bacillus was very much the protagonist of early medical bacteriology, and was also responsible for the first uses of clay filters. The first steps seem to have been taken in Edwin Klebs' laboratories in Berne, during the period he was to look back on as the happiest of his stormy career (see 380). Klebs was involved in the early stages of many of the fundamental discoveries in bacteriology. Whether he was deprived of the satisfaction of making a definitive discovery only by an unkind fate, or whether he had an unfortunate tendency always to make a good start, with a promising idea, only never quite to follow it through, is a matter for conjecture. Certainly his near misses in the fields of tuberculosis, vaccinia, goitre, syphilis and rinderpest outweigh the achievements to which his name is linked, namely the description of *Corynebacterium diphtheriae*, or the Klebs–Loeffler bacillus, and *Klebsiella pneumoniae*. From the history of the development of the bacteria-proof filters his name has also all but disappeared; but there can be little doubt that his was the initiative responsible for the construction of the first clay filters to be used in bacteriological work.

It was towards the end of the Franco-Prussian war of 1870–71 that Klebs suggested to his assistants Tiegel and Zahn the preparation of porous filters of white unbaked clay as a means of separating pathogenic bacteria from the liquids in which they were suspended. It coincided with an attempt to put Dunant's lessons of Solferino, the foundation of the Red Cross, into practice within Swiss territory; an attempt which, paradoxically, showed Klebs, whose personal relations with his staff and students deteriorated disastrously during his later career, at his humanitarian best in caring for ill and exhausted retreating French troops under difficult conditions in an improvised field hospital outside Berne. Here he also found time to pursue his consuming interest in searching for pathogenic micro-organisms in the rich material offered, using for separation of the fluids the method developed in the meanwhile by Tiegel and Zahn, viz. filtration through unbaked white clay cells connected to a Bunsen air pump (446).

Six years later Pasteur and Joubert obtained the final conclusive results of the long saga of anthrax studies using a plaster of Paris filter connected to a vacuum pump (353). In subsequent years, porcelain vases were introduced in the laboratories of Pasteur, and such filters found increasing use when it was realized that they could form a basis for water purification. Chamberland and E. Roux moved from an initial exercise with a broken clay pipe[1] (purchased from the tobacconist supplying Chamberland, who was an inveterate smoker), to vases of incompletely fired porcelain around 1884. In Germany in the last decade of the nineteenth century Nordtmeyer used diato-

---

[1] Still to be seen in one of the treasure-packed cabinets of the Museum of the Institut Pasteur in Paris.

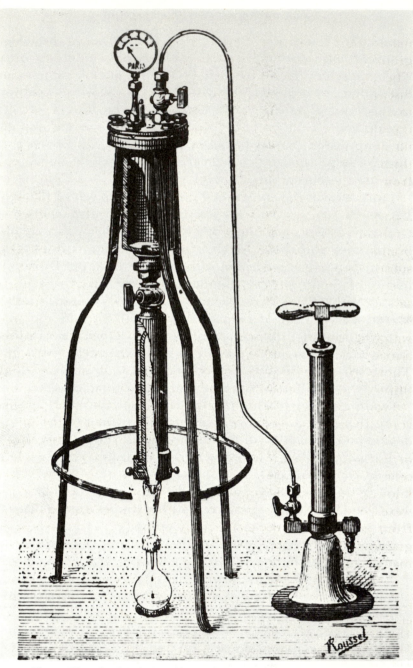

Fig. 5. Chamberland-type filter connected to hand-pump, as used at the Institut Pasteur towards the end of the nineteenth century. From Duclaux (1898), *Traité de microbiologie*, Paris.

maceous earth (*Kieselguhr*) from the Berkefeld mines, where the ground water had been observed to be particularly clear (331). The dates are critical. When serious studies of the mosaic disease of tobacco first began, the use of clay filters was restricted to a small group of pathologists working mainly on anthrax. By the time Ivanovski entered the field, water sterilization by filtration had become a generally accepted procedure, largely through the work of Pasteur and his associates, and filtrates from 'bacteria-proof' filters were expected to be 'sterile'.

The development of filters used by pathologists proceeded in a parallel fashion to that of filters used by chemists. Following Graham's tube experiments, Fick in 1855 used collodion membranes (153) in diffusion studies in an attempt to find reliable substitutes for the animal membranes hitherto employed, and five years later Schumacher described the first simple collodion sacs (406). The two types of filter came together in 1896 when Martin used Pasteur–Chamberland filter candles impregnated with gelatin or silicic acid (308), and pressures of 40 to 50 atmospheres, to separate colloids and crystalloids (as defined by Thomas Graham 35 years earlier according to their diffusibility through his original membrane filters).

From the late nineteenth century onwards, filters underwent continual improvement. It soon became apparent that it was desirable and possible to prepare bacteriological filters of more or less well-defined varying pore size, to yield information on relative sizes of the rapidly growing number of filterable viruses known. Both Pasteur–Chamberland- and Berkefeld-type filters were commercially available in different standard pore sizes from an early date. Berkefeld filters were supplied in three grades marked W, N and V. According to Stuart Mudd,[1] the manufacturers explained the grading in the following way.

The grading of the cylinders, N (normal), V (coarse) and W (dense),* corresponds with a scale showing the filtering capacity with pure (filtered) water under a certain pressure.
The size of the pores is about 3–4 μ in the W grade.
The size of the pores is about 5–7 μ in the N grade.
The size of the pores is about 8–12 μ in the V grade.

[1] This guide to filter grading appeared in Mudd's chapter on filters and filtration in the classical text *Filterable viruses* edited by T. M. Rivers and first published in 1928 (370).

The substance of the filtering cylinders is purely siliceous, and therefore quite indifferent to water, alcoholic, aetherial or organic liquids.

*W signifies '*w*enig Wasser durchlassende Filterzylinder'.
V signifies '*v*iel mehr Wasser als *n*ormale durchlassende Filterzylinder'.

But it was also obvious, especially in studies on the larger viruses, such as those of rabies and vaccinia, that the use of filters presented additional problems. The clogging of pores could seriously affect results; analogies (false as it turned out) with bacteria and protozoa led to apprehension lest viruses might grow through filters when filtering time was extended.

A significant advance was made in 1907, when Bigelow and Gemberling showed the way to obtaining collodion membranes of constant and uniform thickness (44). Even more important, it was also the year in which appeared the first of Bechhold's classic series of papers on the process he named 'ultrafiltration' (32). Bechhold impregnated filter paper with gelatin or acetic acid collodion, and found that he could obtain a range of filters of graded permeabilities by varying the concentration of the solutions used. He also established a method of determining pore sizes of his ultrafilter membranes, and laid the foundations for the development, by himself, and later by Elford (138), of the method for determining the relative sizes of viruses. In the late 1920s and early 1930s the method yielded copious amounts of information on the subject until, after 1935, Pandora's box was opened, revealing more relevant and realistic truths about the nature of viruses; but many of the data recorded in those early days have proved remarkably accurate when judged by the standards later obtaining after the introduction of the electron microscope.

Hand in hand with techniques for filtering virus suspensions went the use of centrifuges. In combination with filtering procedures, they could to some extent circumvent the problem of pore-clogging. The mechanical centrifuge was invented by de Laval for use as a cream separator about the time that filter candles came into use in bacteriological studies.[1] In 1898, Sana-

---

[1] The first centrifuge was designed as a cream separator by the Swedish engineer de Laval, perhaps better known for his invention of the steam turbine, about 1880. Some early centrifuges or cream separators, were horse-driven, but as the development of centrifuges coincided with the establishment of power stations in the Western world, electricity soon took over. De Laval's career is described in *Gustav de Laval 1845–1913* which also contains reproductions of early design sketches by de Laval (100).

relli, who was unable to separate the large poxvirus of rabbit myxomatosis from accompanying tissue elements by filtration, used centrifugation to obtain an 'optically completely pure and totally sterile serum' (392), which nevertheless retained full infectivity. In the early years of the twentieth century, centrifugation was frequently used in conjunction with filtration in

Fig. 6. Early horse-driven centrifuge (note 1, p. 18), from *Gustav de Laval 1845–1913* (100). (Reproduced by courtesy of Tekniska Museet, Stockholm.)

studies on the aetiology of diseases thought to be viral. It was in work on the virus of rabies, by Remlinger and others, that this proved an especially valuable tool, since the suspension of brain and nervous tissue containing the large virus made filtration exceedingly difficult. The speeds of the early centrifuges were modest, usually ranging from 2000 to 5000 rpm. The impact of the invention of the ultracentrifuge by The. Svedberg and R. Fåhraeus (435), after which speeds of 15000 rpm and later as much as 40000 to 60000 rpm became possible, was considerable for virus research. It was preceded by the introduction of differential centrifugation in 1922 by MacCallum and Oppenheimer (297). They used fluids of different specific gravities for suspension of vaccinia virus, and were able to concentrate the virus by stepwise re-suspension and centrifugation of the solutions of known specific gravity. In the 1930s the method was improved with the advent of high-speed centrifuges.

No sooner had it been discovered that the members of a group of disease agents were too small to be retained by bacteriological filters, than it was realized that they possessed one other equally obvious, equally negative, common characteristic: they could not be made visible in the microscopes of the day. The limitations of the available microscopes had been well publicized in papers by Abbé (1), Helmholz (207) and Rayleigh (362), which had made it clear that the limit of resolution was dependent on the wavelength of light used. Early attempts to extend the limit of resolution included the use of ultraviolet light, as early as 1904 (249); later improvements added staining of the material with fluorescent dyes (e.g. primulin).

Long before this, more conventional methods of staining had been employed in virus research. The earliest recorded staining of virus-infected material pre-dates the filterable virus concept by several years. Staining techniques had been employed in histological work from around 1850, beginning with carmine preparations and progressing via haematoxylin, extracted from logwood, to fuchsin and aniline dyes.[1] Early bacteriologists were quick to adapt staining methods for use with bacteria, and part of Robert Koch's outstanding success as a bacteriologist could be ascribed to his flair for developing suitable bacterial stains, and using them in conjunction with the methods he developed for microphotography (247). By the 1880s, many pathologists were actively searching for bacteria in a variety of tissues and tissue fluids. Lymph from vaccinial and variolar vesicles was a favourite object. It was frequently contaminated, and consequently a number of workers in the field found and described bacteria in the lymph, wrongly associating them with the aetiology of the disease.[2]

---

[1] As in much early work when communications were slow and difficult, the history of staining techniques is fraught with claims and counterclaims for priority (90).
[2] As late as 1898, S. M. Copeman grew the 'suppositious micro-organism' of vaccinia lymph in eggs (cf. p. 39).

The methodological background

In 1886, John Brown Buist communicated to the Royal Society of Edinburgh a paper on 'The life-history of the micro-organisms associated with variola and vaccinia. . .'. Buist successfully devised a method for fixing and staining pustular lymph, enabling him to see, and even to measure the size of, what were un-

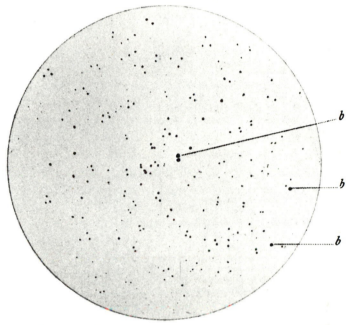

Fig. 7. Buist's stained 'spores of micrococci' from vaccinia lymph in 1886 (56). (Reproduced by courtesy of the Royal Society of Medicine.)

doubtedly the elementary bodies of vaccinia and variola. Although contamination of his material, combined with his preconceptions regarding bacteria and bacterial spores, caused him to draw the wrong conclusions from his observations, Buist should be remembered as the first to have observed virus elementary bodies in fixed and stained material (Fig. 7). Nearly 20 years earlier, in 1868, Keber had filtered vaccinia lymph through 'Swedish filter paper' and found that the filtrate retained infectivity. By microscopic examination of the infectious filtrate he detected 'a number of nuclei and molecules' of 'almost immeasurable tenuity'. He did not stain his material, and published no illustrations, although he had apparently deposited some with the editor of *Virchows Archiv für pathologische Anatomie und Physiologie und für klinische Medizin* (241).

Once viruses had become recognized as filterable and invisible entities, renewed efforts were made to make them visible by staining, and, not surprisingly, a measure of success was obtained with those viruses which form inclusion bodies. They were a particular interest of von Prowazek who, in the first decade of the twentieth century, did a prodigious amount of work on trachoma, fowl plague, vaccinia and some of the animal poxviruses. Von Prowazek was a dedicated microscopist who used the now well-established Giemsa stain, originally created for studies with the malaria agent, to good advantage on trachoma inclusions: well-defined differential staining was obtained, with reddish granules surrounded by a blue substance. Unfortunately, von Prowazek's resulting theory of these structures as 'chlamydozoa' (464), or mantled animals, and the organisms responsible for the associated diseases, was less clearcut and posed more problems than it solved. Little additional progress was made in the field of staining of virus particles until the advent of the electron microscope and the techniques of metal shadowing and negative staining with phosphotungstic acid nearly 50 years later (194).

3. Discovery in the closing years of the nineteenth century

The perfection of the Pasteur–Chamberland filter candle in the mid-1880s and its general acceptance for microbiological studies sparked off the realization that some pathogens, contrary to all expectation, were able to pass through the pores of the filter candles. In view of the amounts of distinguished work done in the very laboratories where the filters were developed, it might seem surprising that rabies was not the first, or even the second or third, virus disease whose agent was labelled 'filterable'. Pasteur himself had made many unsuccessful attempts to isolate its pathogen, and had indeed speculated that it might elude him because of its minute proportions.[1] But here, as in many of its other aspects, rabies presents special difficulties. It is a relatively large virus (approximately 180×80 nm), and the presence of brain and nervous tissue in suspensions hampered filtration.

There were no such difficulties in the cases of the viruses of the mosaic disease of the tobacco plant, and foot-and-mouth disease. By virtue of their small size (although its minimal infective length is about 300 nm, tobacco mosaic virus has a diameter of only 18 nm; the diameter of foot-and-mouth disease virus is 24 nm) and the comparatively uncomplicated fluids in which they were found, they emerged, little more than a decade after the appearance of effective bacteriological filters, as the first 'filterable' viruses on record.

For all the importance and interest afforded by the historical development of ideas on many of the viruses pathogenic to man,

[1] In the early years of the twentieth century, after Pasteur's death and when 'filterable viruses' had emerged as a concept, a number of Pasteur's followers claimed that he had suggested that the pathogen of rabies might be 'invisible'. There is no record of such a statement anywhere in *Oeuvres de Pasteur* (349), although on one occasion his words could perhaps convey this meaning (cf. p. 48).

23

it is a plant virus which has again and again occupied a central position for the solution of many problems in virology. The agent of the mosaic disease of the tobacco plant was the first pathogen to be shown to pass through filter candles; the first to be obtained in pure crystalline form; the first to be identified as infective nucleic acid.[1] Apart from the relative ease and convenience, plant viruses can always claim the advantage that no ethical misgivings need attach to experimental work. On the other hand, the ready availability of, and ease of work with, whole plants have meant that there has been less incentive to develop good assays, let alone a plaque method or tissue culture techniques comparable to those facilitating work with animal viruses (cf. Chapter 6).

Tobacco became established as a crop plant in Europe after the discovery of America, where it had long been cultivated by the American Indians. By the mid-nineteenth century it had become a major crop in Holland and in parts of Russia, when concern arose over the effect on yields of a rapidly spreading disease of the tobacco plant. It was first recorded by a Dutch student working on a tobacco farm during the university holidays, in 1857.[2] The problem grew, damage to crops increased and by 1879 the concern of Dutch tobacco farmers had grown into alarm. They requested help from the authorities, and an official enquiry was launched. The man in charge was the German-born director of the Agricultural Experiment Station at Wageningen, Adolf Mayer.[3] He approached the problem without the preconceptions of some German botanists of the time, who refused to believe that plants were at all susceptible to bacterial attack, and claimed that all plant disease was fungal in origin.[4]

[1] Final evidence was not obtained until 1956 (p. 125) although Schramm had been wrestling with the problem of tobacco mosaic virus since the early years of World War II. Four years earlier in 1952, Hershey and Chase had been able to show that only the nucleic acid moiety of bacteriophage T2 entered the host cell (p. 108).

[2] The student, J. H. van Swieten, wrote a short paper on his observations, but apparently tobacco farmers had by then been familiar with the disease for a number of years (see 313).

[3] Born and educated in Germany, Adolf Mayer spent nearly 30 years, from 1876 to 1904, at Wageningen before returning to live out a ripe old age in his native country (235).

[4] This view was reiterated as late as 1882 in a popular textbook by Hartig, who believed that the cellulose walls and the acidity of their contents provided plant cells with an effective defence against bacterial invasion (200).

Adolf Mayer proved that the disease was transmissible, and he was more than willing to believe in a bacterial origin of the disease, but he also considered the possibility that an 'unorganized ferment' might be responsible. In an effort to elucidate this point, he passed the infectious sap through a double layer of filter paper. Thinking the agent was retained by the filter paper, he concluded that it was bacterial, without finding any bacteria by microscopic examination of the residue. After coining the name of tobacco mosaic disease under which it was to be known from then on (313), Mayer abandoned a problem he felt he could not further clarify.

His filter experiments, and the fact that he had been unable to find an agent visible under the microscope, did not go unnoticed at a time when isolation and characterization of specific pathogens of diseases of animals and man were pursued by an increasing number of pathologists. The year 1879,[1] when Adolf Mayer first undertook to study this plant disease, was also the year in which Paul Bert first examined filtrates from suspensions of brain material from rabid animals passed through plaster of Paris.[2] His results, brought to the notice of Pasteur, were as inconclusive as Mayer's. The filters, which had yielded so many organisms to the scrutiny of Koch and Pasteur, revealed nothing in the cases of tobacco mosaic disease and rabies.

In Russia, Mayer's report was read by the young D. I. Ivanovski, who with a fellow student had been sent to the Ukraine and Bessarabia to report on diseases causing considerable damage to tobacco crops. Having identified one disease as the mosaic disease described by Mayer, he improved on Mayer's results by showing that infectivity of sap from diseased plants was retained after passage through the filter candles developed by Chamberland in 1884 (76). When he published his results in 1892 (227), he believed, as had Mayer, that the disease was of bacterial aetiology; but being unable to isolate and cult-

---

[1] In 1879, bacteriology was a new and promising discipline. The epoch-making results of Koch and of Pasteur had made it a respectable academic subject, and courses in bacteriology were becoming established in schools of pathology at various European universities (474).

[2] This paper (43a) was a short and isolated excursion into the study of rabies on the part of the physiologist Paul Bert before fulltime politics claimed him in the French nineteenth-century tradition.

Fig. 8. Dmitri Iosifovich Ivanovski (1864–1920).

ivate a microbe, he was inclined to identify the pathogen as a toxin-producing entity. This was a logical thought in 1892; four years earlier, Roux and Yersin had discovered the bacterial toxin of diphtheria (386).

Ivanovski took no firm stand on the aetiology of tobacco

mosaic disease. On the basis of his limited experiments, he could hardly have done so. But although Adolf Mayer had abandoned the tobacco plant and its disease after 1886, M. W. Beijerinck, who had watched some of Mayer's initial experiments at Wageningen, had been unable to put the problem out of his mind. Beijerinck, one of the work-obsessed loners of early microbiology, had one enduring friend from his student days. Informal discussions with Jan van't Hoff, who in his early twenties made history in stereochemistry, had sown the seeds of a biochemical approach in Beijerinck's fertile mind.[1] When he had elaborated on Mayer's original experiments, he could not only present surprising results; he also offered a tentative explanation which startled his contemporaries, and even the generation of their successors, to the point of disbelief. Ivanovski, nursing his own bacterial theory, had no use for Beijerinck's radically different approach. Reading other papers on tobacco mosaic disease dating from the early years by Koning (250), Heintzel (203) and a little later Hunger (223), one gets the impression that not only they but even Adolf Mayer found it difficult to follow Beijerinck's flight of thought. E. Roux, and with him the staff of the Institut Pasteur, dismissed Beijerinck's theory in one sentence as 'interesting, but unproven' (385). Loeffler and Frosch, in spite of the tenuity of foot-and-mouth disease virus, continued to be convinced of its microbial nature.

Beijerinck showed, as Ivanovski had done, that the causative agent of tobacco mosaic disease passed easily through porcelain filters. He went on to show that it multiplied in infected tissues, and therefore could not be a toxin; and furthermore, that it could not be cultivated *in vitro* because it multiplied only in living, and preferably growing, tissues (36). Summoning his resources of chemical knowledge, he drew a conclusion which was to remain controversial until techniques and microscopes had been improved sufficiently to vindicate a reasoning which at the time seemed wildly speculative to all but the most unorthodox of microbiologists. Centanni, who made history with fowl plague virus three years later (p. 38), was willing to con-

---

[1] In their impecunious student days at the Delft Polytechnic, van't Hoff and Beijerinck roomed together, and Beijerinck was an enthusiastic helper with some of van't Hoff's early experiments (see 120).

sider 'elements belonging in some transitional area' (74) between living organisms and chemical molecules, but was, with
many another author, disturbed by Beijerinck's insistence on the
'fluid' state of the agent. Sanfelice, some years later, developed
ideas on the pathogenesis of fowl plague (p. 46) which were not
too far removed from Beijerinck's basic concept. But Sanfelice
was not in the habit of making many references to other people's
work, unless he violently disagreed with them; we have no way
of knowing whether he was aware of Beijerinck's papers on
tobacco mosaic disease.

Beijerinck labelled the agent a '*Contagium vivum fluidum*'. It
was not a felicitous choice,[1] and his critics, including Ivanovski,
pounced on this expression and all but ignored the real issue,
which was buried in long and complex German sentences. To
understand the attitudes of both sides, it is necessary to consider
the primitive state of protein chemistry at the turn of the
century. The concept of macromolecules was not to be introduced for another 25 years; even then it remained unacceptable
to most chemists for a long time.[2] In 1898, substances were
considered to be either 'corpuscular', or 'particulate', suspended in the surrounding fluids, like bacteria and blood cells; or
they were 'dissolved', or 'soluble', molecules of low molecular
weight. Beijerinck's stated aim was to 'answer the question of
whether the virus is to be regarded as dissolved or corpuscular'.
Since it passed through bacteria-proof filters, he concluded, it
must be not 'corpuscular'[3] but 'dissolved' and hence 'molecular'. He then showed his mettle in drawing his bold conclusion. The tenor of his considerations remains astonishingly close

[1] Beijerinck's insistence on a non-cellular infectious entity also re-opened the nineteenth-
century debate as to the existence of living entities smaller than the cell, which
involved concepts such as bioplasts (Beale) and microzyma (Béchamp). While C.
Oppenheimer offered some support for Beijerinck's views (338), Ernst Joest dismissed
the *contagium vivum fluidum* in no uncertain terms (234).

[2] In 1925, three years after he had first introduced his macromolecule concept,
Staudinger was all but laughed off the platform at the Zürich Chemical Society when
he expounded his views; and that was by no means the end of resistance to his
ideas (334).

[3] Commenting on Beijerinck's insistence on the 'fluid' rather than the 'corpuscular'
state of the virus, Bawden wrote many years later: '...this suggestion also had its
repercussions, for whether or not viruses were particulate was solemnly debated,
without anyone apparently asking what they could be if they were not particles of
some size or other' (29).

Fig. 9. Martinus Willem Beijerinck (1851–1931).

to the modern concept of the virus. He saw it as a, possibly water-soluble, molecule, able to replicate, but only when 'incorporated into the living protoplasm of the cell, into whose reproduction it is, in a manner of speaking, passively drawn'. It was a momentous pronouncement; and most pathologists, if they noticed it at all, found it very hard to accept.

After the publication of Beijerinck's paper, Ivanovski claimed priority for the filter experiments (228), and Beijerinck immediately acknowledged this (37); but their interpretations continued to be wide apart. In a last paper on the subject,[1] published in 1903 and containing results from his doctoral thesis, Ivanovski reiterated his bacterial theory, and believed that he was on the point of developing a culture method for the supposed bacteria of tobacco mosaic disease (229). His theory was wrong, but his microscopic observations were above reproach and have easily stood the test of time (Fig. 10). They show both crystals of the virus and the mysterious inclusion bodies which puzzled many a later investigator.

In 1897, the year before the publication of Beijerinck's original paper in Dutch (36), the German government had established a commission to investigate outbreaks of foot-and-mouth disease which formed a continual threat to the country's cattle industry. The commission was headed by Friedrich Loeffler and Paul Frosch from Koch's Institut für Infektionskrankheiten in Berlin. They made a very thorough and admirable study of the disease and its aetiology (278; 279). If one is tempted to suggest a causal relationship between the discovery of diphtheria toxin and Ivanovski's filter experiments, one could also see the findings of Loeffler and Frosch as a corollary of Behring's discovery in 1890 of anti-toxic immunity (35). The avowed aim of the commission was to prevent further outbreaks, perhaps by producing a vaccine. It was when they inoculated healthy calves with filtered vesicle lymph which they hoped might contain anti-toxin and so confer passive immunity that Loeffler and Frosch found the calves developing characteristic symptoms of the disease. By extensive serial inoculations and re-inoculations with in-

---

[1] Ivanovski never returned to the subject of plant virus disease after 1903, but turned to plant physiology on which subject he wrote extensively until his early death in 1920 at the age of 56 (261).

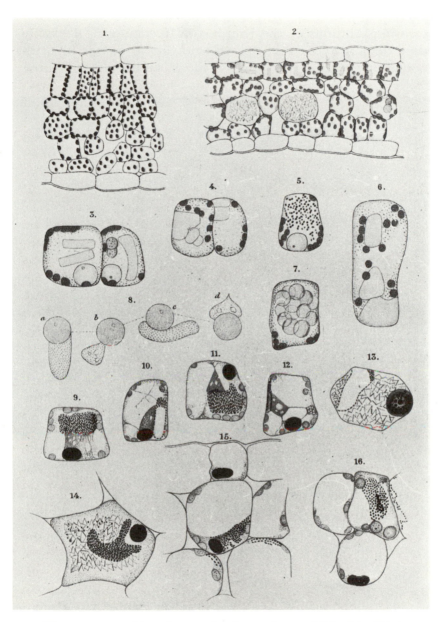

Fig. 10. Ivanovski's microscopic observations published in *Zeitschrift für Pflanzenkrankheiten* in 1903 (229).

creasing dilutions of the lymph, and meticulous calculations, they demonstrated that this was no simple matter of toxins. Something passed through the filter candles which had the

Fig. 11. Friedrich Loeffler (right) and his great teacher and mentor, Robert Koch.

capacity to multiply in the infected animals. Loeffler and Frosch, true disciples of Koch, drew the inevitable conclusion:

The agents must then be so small that they are able to pass through the pores of a filter which will certainly retain even the smallest bacteria. The smallest bacteria hitherto known are the influenza bacilli described by Pfeiffer. Their length is 0.5–1.0 μ. If the supposed agents of foot-and-mouth disease were only one-tenth to one-fifth of this size, which is by no means impossible, then, according to the calculation of Professor Abbé of Jena on the limit of resolution of our microscope, they would be unrecognizable even by the best of modern oil-immersion systems.

Shortly afterwards, pathologists at the Institut Pasteur in Paris were investigating another cattle disease, namely bovine pleuropneumonia. Its agent also passed through filter candles; but it could be seen, if only as tiny points, in the light microscope, and it could be grown in special collodion sacs implanted in the peritoneum of live guinea pigs (329). We know now that it is a mycoplasma, differing from viruses by containing both RNA and DNA, by being able to grow on artificial media, and by multiplying by binary fission. But in 1898, it was immediately hailed as a filterable virus, and thereby caused considerable confusion during the early years of the concept. It encouraged pathologists to believe that it would be possible in time to develop methods of growing other filterable viruses if the right conditions could be found, and even to make them visible, perhaps by staining, or by improvements in the resolution of microscopes.

A fourth pathogen was tentatively included in the group before the turn of the century. Giuseppe Sanarelli, discoverer of the haemorrhagic allergy which bears his name, was a disciple of Metchnikoff and throughout his life kept up his ties with the Institut Pasteur while holding posts at universities in his native Italy. In 1896 he was invited by the University of Uruguay to establish and direct an institute for experimental hygiene at Montevideo. Here he found his work on yellow fever (391) interrupted by an outbreak of a frequently fatal disease among his laboratory rabbits.[1] In spite of his efforts, he could detect no microbe in the lymph from the vesicles of the dead and dying rabbits, and he described the virus of myxomatosis as invisible (392). Not surprisingly, he was initially unable to demonstrate its filterability; this large poxvirus measures approximately 285×230 nm, or roughly 10 times the dimensions of foot-and-mouth disease virus.

To sum up the position of the concept of the virus as a separate biological entity at the turn of the century, it had by then emerged from its background of general bacteriology in a tentative way. A total of four infectious agents were accepted as belonging in a group variously referred to as 'filterable' or 'invisible' viruses. In addition to what we might call the two

---

[1] It was an early striking example of the effects of transmission of a virus endemic in one species (the tropical forest rabbit of South America *Sylvilagus brasiliensis*) in which it produced a mild or even inapparent infection, to a related species (the European domestic rabbit *Oryctolagus cuniculus*) which had not previously encountered the virus, and consequently responded strongly, with almost 100% fatality (see 152 and 65).

pioneer viruses of tobacco mosaic and foot-and-mouth disease, they included the agent of bovine pleuropneumonia and that of rabbit myxomatosis.

The agents of tobacco mosaic disease and of foot-and-mouth disease had been shown unequivocally to pass through filter candles and to reproduce the respective syndromes from clear, bacteriologically sterile filtrates. Their discoverers, although far from agreeing with each other, had in their extensive studies and their different interpretations laid down the guidelines for the discussions on the mysterious nature of these pathogens for many years to come. Beijerinck's boldly imaginative concept of tobacco mosaic virus as a molecule dissolved in the juice of the infected plant and endowed with the power of replication, but only with the help of the existing mechanism of the infected cell, was in sharp contrast to the more conservative view held by Loeffler and Frosch at the conclusion of their work on foot-and-mouth disease, i.e. that its agent was a micro-organism of bacterial nature but of extreme tenuity.

On the other hand, the agent of bovine pleuropneumonia was destined to contribute nothing but confusion to the whole concept of filterable viruses for half a century. Visible under the light microscope, if only just, and growing in artificial culture, although under rather special conditions, its example encouraged those who believed, as did many pathologists at the time, that the failure to observe viruses with the aid of the light microscope and to grow them *in vitro* represented only temporary obstacles which would be overcome as techniques improved. In this way the pleuropneumonia organism continued to confuse the issue until it was finally characterized as a mycoplasma more than 50 years later (134). The myxomatosis agent, although undeniably a virus, never assumed such general importance. The most highly specific of viruses – no other mammalian host is known, even the hare being immune – it is a large poxvirus. Filtration experiments were consequently difficult, and Sanarelli initially described it as an agent 'beyond the visible'.

4. Filterability and the nature of filterable agents

From the turn of the century until the outbreak of World War I there was a great deal of activity in the field of filterable viruses. Much of it was directed simply towards proving the filterability of those disease agents which had so far eluded recognition by conventional bacteriological methods. An equal amount of energy was spent in the search for artificial media which would allow the culture of the mysterious entities without the time-consuming and costly use of animal experiments. Of the vast amount of literature on a multitude of virus diseases accumulated over this period, some deserve more particular attention. By virtue of the approach used, the results obtained, and the men who obtained them, they stand out from the rest. The first such case to heighten our interest can be found within the first few days of the new century.

African horse sickness

In January 1900 John M'Fadyean of the Royal Veterinary College, who in 1888 had founded the *Journal of Comparative Pathology and Therapy*, published as the first paper of its thirteenth volume an article on 'African horse-sickness' (299). M'Fadyean showed the agent to be filterable, using samples of infected horse blood brought back to London by an English veterinarian.[1] The continued infectivity of the serum passed through Berkefeld and Chamberland F filters after the long voyage back from

---

[1] One W. Robertson, M.R.C.V.S., who, with the director of the Cape Bacteriological Institute, had gone to Kimberley with Robert Koch in December, 1896, when Koch's objective was the development of a vaccine against rinderpest.

35

South Africa[1] demonstrated not only the ability of the pathogen to pass through bacteria-proof filters, but also the remarkable stability of some filter-passing viruses.

It may seem surprising that an indigenous African virus disease of horses should be first described as filterable in London. It had in fact been under study in South Africa since 1887. In May 1901, the *Deutsche Thierärztliche Wochenschrift* carried an article entitled 'Die südafrikanische Pferdesterbe' (440). Its author was Arnold Theiler, father of Max Theiler who many years later was to receive a Nobel Prize for his contributions to the conquest of another virus disease, namely yellow fever. Arnold Theiler had been working on African horse sickness as government veterinarian in Pretoria for nearly 10 years by the time M'Fadyean's paper appeared,[2] and had had as his assistant the young Jules Bordet. He also had found the pathogen to pass through filter candles. Theiler continued to work on the disease, and gradually perfected a technique of immunization of horses which consisted of simultaneous intrajugular injection of immune serum and live virus from diseased animals. In the course of developing this technique Theiler in 1918 recorded an outbreak of serum hepatitis in immunized horses, at a time when the analogous condition in man had received scant attention.[3] He expressed no clear views on the aetiology of the disease; his primary concern was with African horse sickness, and the development of a vaccine. On its own terms, it is a very full account (441), and even includes what appears to be a case of sexual transmission of serum hepatitis in the horse.

M'Fadyean did not in that first paper on horse sickness speculate on the nature of the agent involved. He referred to it as a 'microbe' and considered that it did not 'attach itself to the red corpuscles of the blood', a remark which offers the first mention in virus research of an affinity which, where it does

---

[1] The serum samples travelled, protected by a layer of glycerol, in sealed glass tubes; glycerol had long been an accepted preservative for bacteriological samples.

[2] There is a note of veiled regret at having missed priority of publication in Theiler's introductory paragraph when he tells his readers that after working on the subject for nearly 10 years, and having filtered the pathogen through Chamberland candles, he 'happened to notice' M'Fadyean's paper while preparing his own for publication.

[3] The first full report of what was clearly an outbreak of serum hepatitis in man is found in a paper by Lürman dated 1885 (289), when the virus was accidentally transmitted with smallpox vaccination lymph to dockyard workers in Bremen (p. 170) in 1883.

exist, as in the case of the influenza viruses, was subsequently to prove of enormous importance. Hence, when the virus of African horse sickness joined the list of pathogens which had been proved to pass through bacteria-proof filter candles in the first days of the twentieth century, Beijerinck had gained no support for his views on the nature of the virus, although it is evident from all the contemporary writings that his work and his theories were well known to pathologists whether they worked with viruses of plants, animals or man.

It is also clear from the literature of the period that by this time any disease agent which could not immediately be identified in the light microscope and grown by the now well-established culture methods so recently developed by Koch and his school was suspected of being filterable. Filter candles of various types were in great demand, and improvements in the technique were frequently proposed and discussed, especially by members of the French school working in and around the Institut Pasteur in Paris. They were by now convinced that the rabies agent must also belong in this category in spite of the difficulty of isolating the pathogen from the filter-clogging suspensions of brain and other nervous tissue.

Fowl plague

Before the problems surrounding the passage of the pathogen of rabies through filter pores had been solved, another virus which was later to prove of even greater general importance was making its presence felt in Europe. Fowl plague had been known and described in Italy during the latter half of the nineteenth century; the difference in aetiology became clearly recognized only after Perroncito in Turin, and Pasteur in Paris, solved many of the problems surrounding chicken cholera between 1878 and 1880. In the spring of 1901 there was an outbreak of fowl plague in Italy and, as no causal organism had been isolated, pathologists tried, and succeeded, in passing the infective principle through filter candles. The first to prove its filterability was Eugenio Centanni (75), then professor of pathology at Ferrara, along with E. Savonuzzi; but a number of papers (see 476) published in rapid succession map the progress of the epizootic through Italy and across the Alps into Austria with the stock of an itinerant poultry merchant and finally into the German states via the Brunswick poultry show in the summer of 1901.[1]

At first sight fowl plague might appear to be just another animal virus disease, of little interest save to poultry farmers and veterinarians actively concerned with the survival of the poultry industry; and yet, for a variety of reasons, its virus has attracted considerable attention among virologists from the time it was first isolated in 1901 until the present day. Centanni, whose priority for isolation of fowl plague and demonstration of its filterability was tenuous but undisputed, pointed out as early as 1902 that this pathogen seemed ideally suited as a model for the study of the nature and properties of viruses in general. His opinion was echoed by a number of early workers in the field of filterable viruses, and a very considerable corpus of studies on various aspects of fowl plague virus accumulated even before the outbreak of World War I. The early bibliography[2] of this pathogen provides not only a broad spectrum of many facets of virology; it also evokes many names of men who in one way or another were to change the face and pattern of medicine between the two world wars. More recently, the emphasis has changed. In 1955, Schäfer in Tübingen, in a comparative sero-immunological study (398), identified fowl plague as an avian influenza A, and suggested that in certain circumstances members of this group of viruses might change their host specificity, and that 'in this way a new type of influenza agent could develop from fowl plague virus and vice versa'. From then on, fowl plague virus could more accurately be described as a model myxovirus; but in 1901 such a sophisticated approach was very far from possible.

Yet even in the very early days Centanni realized the importance of the new virus he had isolated. In 1902 he published a description, 28 pages long, and study of the disease in German (74); and Centanni did not often write in German, preferring

[1] When it was realized that a rapidly spreading fatal infection was abroad, the authorities panicked and closed the show, returning all entrants, and with them the infection, to their home states.
[2] The early literature on fowl plague is discussed more fully in (476).

as a rule his native Italian. The choice of language and journal (*Zentralblatt für Bakteriologie, Parasitenkunde, Infektionskrankheiten und Hygiene*) alike reflect a conscious aim at a wider audience. In addition to a great deal of clinical and histological detail, Centanni described a number of attempts, albeit unsuccessful, to grow the disease agent *in vitro* in a specially constructed U-shaped glass vessel (Fig. 12). Among the substrates he used were chicken bouillon, filtered extract of hen tissues in physiological saline, white of egg and fresh yolk of egg. Having obtained no proliferation of virus by any of these methods, Centanni proceeded to test the infectivity of eggs laid by sick hens, or even taken from hens which had died of fowl plague. Administered orally, mixed with its regular feed over a period of six days, they had no effect on a healthy hen.

By accident or design, or possibly by a felicitous mixture of both, Centanni then hit upon a method which in the end was to prove a very important one, i.e. inoculation of embryonated eggs with virulent fowl plague virus. Judging by his presentation, the problem which led him to experiment with eggs was that of the possible transmission of acquired immunity from the maternal organism to the embryo, a question which had been much to the fore in rabies studies in the late nineteenth century. Three years earlier, S. Monckton Copeman had recorded an attempt to grow the causal organism of smallpox in eggs (92); he was under the impression that he was dealing with a microorganism, and thought he could attenuate it by serial passage in calves, and indeed reported some success when using the resulting lymph to vaccinate a number of children.[1] It is not clear exactly what Copeman was growing in his eggs, or in the calves.

Centanni's egg experiments were not extensive. He inoculated a total of four eggs with infected blood, resealing them after inoculation with paraffin wax and collodion. After 22 days of incubation the eggs were opened and examined. One was badly decomposed, another showed no trace of either an embryo or any deterioration. The remaining two eggs each contained an imperfect embryo, one of them with a manifest deformity in one

[1] In this and subsequent publications Copeman also described the technique of using glycerinated lymph, which was officially adopted at home and abroad from 1898 onwards.

keit, auch wenn sie von Virus strotzt, vollkommen klar bleiben, wie in nicht geimpften Röhren.

Man kann sich nur auf das zweite Kriterium, das der Virulenz, verlassen, und diese kann auf drei Weisen erprobt werden:

1) durch zunehmende Verdünnung des Virus in dem Kulturmittel, indem man diejenigen Verdünnungen anmerkt, die vor der Anstellung der Kultur wirksam waren und es nachher werden;

2) durch mehrfache Uebertragungen in sehr große Gefäße, so daß man bei den Reihenimpfungen kleine Mengen aus dem einen ins andere überträgt und endlich nach zahlreichen wiederholten Uebertragungen das ursprüngliche Material in eine unendlich schwache Verdünnung verwandelt hat;

3) mittels der Diffusion des Wachstums. Dabei benutzt man lange, gebogene Röhren; man impft das eine Ende und untersucht, ob das Virus durch fortschreitende Durchdringung des Nährbodens das andere Ende erreicht.

Ich habe vorzugsweise das letzte Verfahren angewendet, mit Anwendung des Apparates Fig. 1. Es handelt sich um ein einfaches, U-förmiges Rohr, welches an beiden Enden zu einer Kugel aufgeblasen und unten ziemlich dünn auf eine Länge von ungefähr 25 cm ausgezogen ist. Man gießt die Nährflüssigkeit ein bis zur Füllung der Kugeln und impft eine dieser Kugeln A mit virulentem Material. Die Röhre kann man zur anaëroben Kultur an beiden Enden zuschmelzen oder offen lassen.

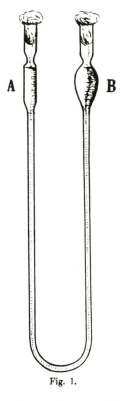

Fig. 1.

Nachdem man die Röhre hinreichende Zeit in der Wärmekammer gehalten hat, aspiriert man einen Teil des Inhaltes der Kugel B (ungeimpft) und erprobt seine Virulenz an einem Tiere. Wenn das Material unwirksam ist, hat sich das Virus nicht vermehrt. Aber wenn das Tier von der Krankheit ergriffen wird, ist der entgegengesetzte Schluß noch nicht berechtigt, denn es kann sich um bloße mechanische Fortführung des Impfstoffes durch Diffusionsbewegung beider Flüssigkeiten handeln.

In unserem Falle sind die Röhren immer mit dem Herzblute des infizierten Huhnes geimpft worden; man hat offene und verschlossene Röhren benutzt, die Dauer der Kultur hat sich auf 7—9 Tage erstreckt und die Temperatur der Wärmekammer ist auf 39—39,5 ⁰ erhöht worden. Die Röhren wurden gefüllt:

a) Mit peptonisierter Fleischbrühe von einem gesunden Huhne;

b) mit Hühnerextrakt, bereitet durch Emulgierung von Geweben in der Kälte mit der physiologischen NaCl-Lösung und Sterilisierung durch die poröse Kerze;

c) mit Martin'scher Fleischbrühe mit Zugabe einer kleinen Menge von Hühnerserum;

Fig. 12. Centanni's U-shaped culture vessel in which he attempted to grow fowl plague virus. From *Zentralblatt für Bakteriologie, Parasitenkunde, Infektionskrankheiten und Hygiene*, 1902 (74).

eye. Thus the effect which was to be developed into the method of titrating fowl plague virus by measuring the rate at which it kills developing embryos in the embryonated egg was recorded initially more than 30 years before it was put to any quantitative use. Centanni merely concluded that the embryo was able to develop in infected eggs but without reaching maturity, and that the virus survived but did not multiply. Had he used a greater number of eggs and pursued this line of thought further and quantitatively, he would have anticipated not only the titration of fowl plague virus, but also the culture of viruses in the chorio-allantoic cells of the developing chick; the latter was rather surprisingly not shown to be applicable to fowl plague virus until after Woodruff and Goodpasture had developed the technique for use with fowl pox virus in 1931 (488; cf. p. 74). Centanni, familiar with Beijerinck's work and ideas, was not prepared to speculate in print to the same extent; yet he kept his options open when he wrote: 'In view of the incomplete methods currently at our disposal, we must of course defer the question of whether the reproduction of this virus involves living organisms or complex chemical molecules, or even elements belonging in some transitional area between the two.'[1]

The Austrians Lode and Gruber, who recorded the outbreak of fowl plague in the Upper Inn valley and pointed out the way in which the disease had travelled across the Alps with the stock of 'a certain Salvatori' (275), an itinerant Italian poultry merchant, were at first tempted to invoke humoral pathology and to propose that the pathogen of fowl plague might have characteristics 'something like an enzyme' (*etwa von enzymartigen Charakter*). After additional filtration experiments, however, Lode decided less than a year later (276) that he could not be dealing with a dissolved substance, and that the only practicable alternative was the 'ultrasmall bacterium' concept of Loeffler and Frosch.

Since Centanni's unsuccessful attempts, the question of whether filterable viruses could be grown on artificial media continued to occupy the minds of pathologists. Marchoux, who is otherwise known for his work on dysentery and leprosy, made

---

[1] What in Centanni's German text is referred to as '...Elemente, die an der Uebergangsgrenze des einen Reichs zum anderen stehen'.

an isolated attempt to grow fowl plague virus *in vitro* (306). He achieved some proliferation of the virus when he added defibrinated blood to glucose–peptone agar in a test tube. He assumed that a zone had been established in which essential chemical substances from the blood had diffused into the agar medium which was then able to support the growth of the virus; he believed that he had thus demonstrated the ability of fowl plague virus to grow on an artificial medium. However, his conclusions were soon questioned by Landsteiner and Berliner (255), who pointed out that in all probability the determining factor was the presence of whole intact blood cells, and that if these, rather than a hypothetical substance released from them, were necessary for the multiplication of the virus, the claim of successful culture on lifeless material was no longer valid.

On the whole, the years until 1915 saw a growing realization of the inability of viruses to grow outside living cells. There was no one dramatic piece of unequivocal evidence, no one presentation of irrefutable fact; just a slow accumulation of suggestive data, confounding those who sought to grow viruses *in vitro*. In 1908, when Marchoux's paper was published in Paris, M'Fadyean had declared in London: 'Another character common to all the ultravisible organisms is that they appear to be obligatory parasites' (300).

Landsteiner and Berliner's paper (255) was published in 1912. Later in the same year Mrowka, a staff veterinarian at the leasehold German naval base at Tsingtao, a major port on the Yellow Sea, made an important contribution to the development of the concept of the virus (315). Mrowka made an attempt to apply the concepts of the young science of colloid chemistry[1] to the even younger notion of filterable viruses. Explaining his intentions, Mrowka wrote: 'The following experiments were planned on the assumption that in the case of the filterable viruses there can be no question of organised elements, whether bacteria or protozoa...the indisputable ability of the filterable virus to pass through capillary walls repeatedly suggested to me

[1] Thomas Graham (p. 14) has been referred to as the father of colloid chemistry (88) with a series of papers published in *Philosophical Transactions* between 1850 and 1864. At the turn of the century, studies by Wolfgang Pauli in Vienna (albumins) and by W. B. Hardy in England (precipitation studies with globulin) and from 1907, Bechhold's filtration studies (p. 18), drew general attention to the subject.

the colloid nature of the virus. Corroborative evidence comes from studies of bacterial antigens and antitoxins which are generally regarded as colloidal in nature.' On the question of proliferation of the virus he wrote: 'Our knowledge of the nature of the virus strongly favours the view that the filterable virus flourishes only in the body of the host and is very probably tied to a specific protein of the living body, being able to multiply only within the living organism.' In Mrowka's experiments the globulins were precipitated out of infective serum by means of tannin, then washed, re-suspended and the suspension subjected to renewed centrifugation. In all cases the globulin fraction remained virulent whereas the supernatant both before and after washing and re-suspension contained no active virus.

Mrowka came to the conclusion that fowl plague virus behaved in all respects as a colloid globulin, and must be regarded as such. For the reader the paper is long and difficult. Much of Mrowka's theorizing appears hazy and even ambiguous, and perhaps for this reason it seems to have had little impact at the time of publication; but comment did come from Andriewsky, who in 1914 made an attempt to determine the size relations of fowl plague virus. This involved comparing the pathogen with haemoglobin in a series of comparative filtration experiments based on Bechhold's ultrafiltration method. Andriewsky concluded that the virus particles must be smaller than the size usually accepted for the haemoglobin molecule,[1] about 2.5 $\mu\mu$ (1 $\mu\mu$ = 1 nm), a very rough approximation based on an empirical formula (15).

Although the work of Mrowka and Andriewsky appears to have been interrupted by the outbreak of the 1914–18 war, and although Mrowka's ideas seem to have received little attention in the contemporary world of pathology, they were not entirely dismissed and forgotten. As we shall see in a following chapter, they were brought out of obscurity in the 1920s to act as a stimulus in an unexpected quarter. When the plant pathologists were searching for biochemical methods which might help them in their efforts to precipitate tobacco mosaic virus from infected

---

[1] A few years later Duggar and Karrer (124) were to point out that Andriewsky appeared to have confused the concepts of molecules and particles (cf. p. 118).

plant juice, they remembered these papers; and when some of their results suggested that the virus reacted to certain treatments in the manner of globulins, they saw in Mrowka's and Andriewsky's results a certain amount of corroborative evidence (cf. p. 121).

Fowl pox

Within a year of the publication of Mrowka's study, Sanfelice in Modena reported experiments which were more precise, and made claims less vague and more provocative, than those of Mrowka. They were based on results with another virus disease affecting the fowl. Sanfelice had been investigating the aetiology of the so-called epithelioma contagiosum or fowl pox in pigeons for nearly 20 years. As in other poxvirus diseases, inclusion bodies are formed in infected tissues, and, as in other cases, the inclusions were identified in various ways, frequently contributing to the confusion.

In some cases, because of their size, inclusion bodies had been observed and described surprisingly early. This is true of the inclusion bodies associated with molluscum contagiosum, which were described by Paterson (354) and by Henderson (208) in the *Edinburgh Medical and Surgical Journal* in 1841. Paterson's meticulous drawings of his observations are reproduced in Fig. 13.

As early as 1897, Sanfelice himself had associated the Bollinger bodies[1] with certain fungi isolated from the skin of infected pigeons (see 393). In 1902, Marx and Sticker had been able to filter the pathogen (see 366), although filter experiments proved positive only with Berkefeld filters, and negative with porcelain ones (Chamberland F). A few years later the histological findings of von Prowazek and of Lipschütz led to their respective postulates of the inclusion bodies as chalmydozoa and strongyloplasma (p. 80). Sanfelice was not much impressed by these theoretical considerations; but the results of staining of the inclusions, where some of the components stained rather like nuclei, suggested to him an approach based on the work of the

---

[1] The inclusion bodies associated with fowl pox were first observed by Rivolta (1869) and by Bollinger (1873). Bollinger named the disease epithelioma contagiosum, but Sanfelice preferred fowl pox (*Geflügelpocken*). In 1897 he had placed the pathogen in the blastomycetes group of fungi (393).

44

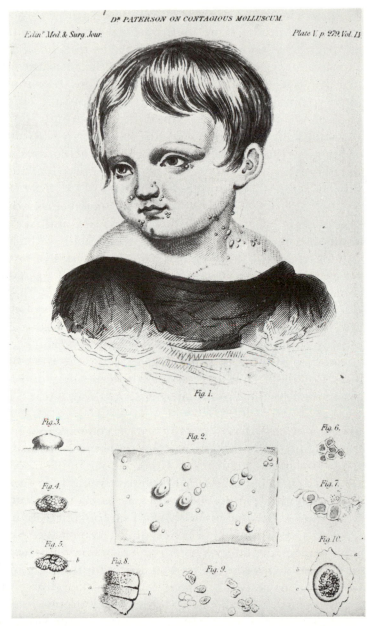

Fig. 13. Plate from Paterson's (1841) paper on molluscum contagiosum, from the *Edinburgh Medical and Surgical Journal*, including
identifiable inclusion bodies (354). (Reproduced by courtesy of the
Royal Society of Medicine.)

physiological chemists.[1] He found that he could extract the active principle by the same method used to extract 'nucleoproteid' (sic), and that it remained infective by the end of rather harsh treatment which would destroy even the resistant spores of anthrax (394). Sanfelice concluded that fowl pox must be caused by a poison in the form of a 'nucleoproteid' formed in the diseased cells, and explained the apparently limitless transmissibility by assuming that the substance was able to induce its own replication by existing cell mechanisms when inoculated into healthy cells. He made no attempt to explain how the 'poison' came to be present in the first place. Years later, in 1927, he returned to the subject. Little had been added to his experimental basis, but his thinking was now influenced by the results obtained in the meantime in the fields of bacteriophages, enzymes and Rous' chicken sarcoma, since part of the interest generated by epithelioma contagiosum had always been the similarity of the lesions to tumours (Bollinger had pointed this out in his paper of 1873). By 1927, Sanfelice concluded that the characteristics of fowl pox virus suggested a closer relationship to the enzymes than to any living micro-organism (395). Sanfelice's view at this time was very close to the working hypothesis of those contemporaries who were engaged in the attempt to identify tobacco mosaic virus, but there is no evidence that he was aware of it.

Sanfelice's broadly based perspicacity is reflected not only in his linking of nucleoproteins to the virus concept at such an early stage, but also in his inclusion of Rous' results in his discussion in 1927. Rous had begun work on transmissible fowl tumours in 1909, shortly after joining the staff of the Rockefeller Institute. When he found that the spindle-celled sarcoma of a Plymouth Rock hen was transmissible in series, and from cell-free filtrates (381), he was surprised and excited. His excitement was not shared by those who represented mainstream thinking in contemporary cancer research, and to whom any suggestion of an infectious origin of cancer was anathema. Rous serenely pursued his own ideas against considerable opposition; his cause received a boost when W. E. Gye began studying the chicken

---

[1] He had already used physico-chemical methods in 1897 in an effort to demonstrate similarities between the agent of fowl pox and known blastomycetes.

sarcomas from 1925 on (cf. p. 159). If the importance of Rous' findings was not recognized for a surprisingly long time, this applied also to a discovery made in 1908 in Denmark, when Ellermann and Bang were able to demonstrate the filterability of the agent of a transmissible fowl leukaemia (140). In those early years there was no evidence to link leukaemias with cancer, and Ellermann's and Bang's discovery received no more than passing recognition as just another filterable virus disease. There was no reason at the time to see any connexion between the filterable virus of chicken leukaemia and Rous' sarcoma agent.

Among the many important results following each other in rapid succession in the early years of the century was the heroic work on yellow fever. For more than two centuries the disease had remained a mysterious scourge in the subtropical areas of the Americas since it first travelled there on slave ships from Africa to Barbados in 1647. The perspicacious Cuban physician Carlos Finlay y Barres had maintained since the early 1880s that yellow fever was probably carried by mosquitoes. The conclusion of the Spanish–American war at the turn of the century became the beginning of a new era in yellow fever research, when Havana was chosen as the site for a United States Army Commission established to investigate the disease.

The Commission was headed by Walter Reed and James Carroll; the dedicated Finlay supplied their initial breeding stock of mosquito eggs. In less than two years the Commission completed its task in an eminent way, proving the validity of Finlay's mosquito theory, and showing that one species of mosquito in particular (*Aëdes aegypti*) was responsible for transmission, a fact which facilitated the task of elimination (364). At the conclusion of the Commission's work, much useful knowledge had been gained, and in addition the agent of yellow fever had emerged as the first pathogen of man to be included in the growing list of filterable viruses;[1] but the cost had been high. In the absence of a suitable animal host for the disease, transmission experiments could be made only with human volun-

---

[1] Reed and Carroll made a point of thanking Wm. H. Welch for drawing their attention to Loeffler and Frosch's foot-and-mouth disease experiments which led them to filter experiments with yellow fever (364).

47

teers. Although none of the soldiers experimentally infected during the work of the Commission died, there had been fatalities among volunteers before the arrival of the Commission. James Carroll himself was ill for a month with experimental yellow fever; Lazear paid the ultimate price for what appears to have been accidental transmission.[1] His life was not the last to be claimed by yellow fever research; more tragedies were to come on the eve of a more thorough understanding of the pathogenesis in the late 1920s (p. 145).

The list of filterable virus diseases was now growing rapidly; Nicolle and Adil-Bey filtered the virus of rinderpest also in 1902 (see 366), and shortly afterwards, Remlinger finally succeeded in passing rabies virus through filters, by judicious use of centrifugation and dilution of suspensions in combination with the largest pore size of Berkefeld filters (365). If documentation of the filterability of rabies virus arrived almost as an after-thought after so many other viruses had been included in the new group, it was fittingly achieved by a disciple of Pasteur. Paul Remlinger, last of the '*vieux pasteuriens*', died in 1964 at the age of 94. Another dedicated recluse, he made rabies his life's work from his early years in Paris and throughout his solitary life as director of the Institut Pasteur at Tangier.[2] When Pasteur had first turned his attention to rabies at the beginning of the 1880s, he had surprised the world by developing a vaccine against a disease whose causal agent he had been unable to isolate by conventional means. The importance of his achievement and the public acclaim it received temporarily obscured the mystery of the identity of the pathogen. Pasteur himself described the agent as '*infiniment petit*' and his friends and colleagues (including Remlinger and E. Roux) after his death – and after 'filterable' and 'invisible' viruses had become an established fact – claimed that he had suggested, in 1881, that the agent might be beyond the limits of visibility altogether. Certainly some of those working in and around Pasteur's laboratories examined the effects of filtration through plaster of Paris on suspensions

[1] Walter Reed died later in the same year the final report was published (1902); but he died of peritonitis following appendicectomy.
[2] According to his obituary 'without family and pupils...devoting all his energies to his work, eschewing academic reunions and even the agitation of scientific conferences' (see 475).

of rabies-infected tissues as early as 1880,[1] more than 10 years before Ivanovski, to his surprise, found that the pathogen of tobacco mosaic disease passed through filter candles. However, there is nowhere in Pasteur's writings any record of his thoughts on the visibility of the rabies virus, apart from his remarks on the microscopic differentiation of rabid and non-rabid brain material: '...one is tempted to believe in a microbe of infinite tenuity, possessing neither the shape of a bacillus, nor that of a constricted micrococcus: they appear as discrete points' (352). There is certainly no evidence at all to suggest that he would have considered it other than a very small microbe, had he realized the extent of its invisibility.

In the same year of 1903, when Remlinger demonstrated the filterability of the rabies virus, Negri described the characteristic and diagnostically important cytoplasmic inclusion bodies which he wrongly assumed to be a protozoal parasite. Two years later Negri made another contribution to filterable virus research when he demonstrated the filterability of vaccinia virus through Berkefeld V filters. Negri then returned to the question of inclusion bodies in rabies, and in a last paper he identified the Negri bodies as a protozoal parasite, *Neuroryctes hydrophobiae* Calkins, and the pathogen of rabies. Three years later he died an untimely death at the age of 36 (323–5).

Activity was at a peak in the field of filterable viruses throughout the first decade of the twentieth century. In 1903, Roux published in *Bulletin de l'Institut Pasteur* a review article on 'invisible' viruses (385). He included a total of nine known viruses. Three years later, Remlinger repeated the exercise, with a slight variation of title, referring to the 'filterable' microbes (366); the number had by then doubled. After yet another two years, in 1908, M'Fadyean, in London, wrote on the 'ultravisible' viruses (300). The titles reflect the uncertainty with regard to classification and characterization which beset early virus research and which was to prevail for a long time. 'Filterable viruses' became the accepted generic term until the virus concept came of age and 'virus' finally acquired a well-defined meaning of its own, very much later. The 'filterable viruses' of

[1] Remlinger in 1903 mentioned that both Nocard, in 1880, and Paul Bert, in 1882, had failed to pass rabies virus through the primitive filters then used (cf. p. 25).

49

the classic textbook edited by T. M. Rivers (370) and first published in 1928 included a variety of disparate infectious agents, some of which are no longer classified as viruses.

Throughout the 1900–10 decade of preoccupation on the part of pathologists with the filterability of disease agents, epidemiologists and paediatricians were increasingly concerned about the growing frequency and extent of outbreaks of what was to become known as poliomyelitis.

When, in 1912, Landsteiner suggested that Marchoux's culture experiments with fowl plague virus might be explained by the presence of whole blood cells, he had already been involved in an exercise which had developed against the background of the impossibility of growing viruses on artificial media. With Popper, he had made an attempt to produce experimental poliomyelitis in animal hosts by injections of spinal cord material from a nine-year-old boy who had died of the disease. Results with rabbits, guinea pigs and mice were negative; but fate took a hand. It so happened that the Vienna pathology laboratories had two monkeys 'left over' from syphilis experiments. J. R. Paul, who had personal experience of the high cost of monkeys as experimental animals, expressed sympathetic concern over the expense incurred by the use of such animals '...which probably represented an expensive luxury for these young scientists' (356) – in this particular case, they seem to have been there for the asking. Nevertheless, the decision to use them proved inspired; and as sometimes happens in the history of scientific ideas, inspiration was sustained by great good fortune. The two monkeys which were available to Landsteiner and Popper were Old World monkeys, readily susceptible to experimental poliomyelitis where New World ones are relatively insusceptible; furthermore, among the Old World ones, the two species accidentally waiting to be used that day in 1908 were among the most highly susceptible. The results were all that could be hoped for: not only the clinical picture, but the lesions in the spinal cords of the two monkeys closely resembled those of the boy from whom the disease had been transmitted (258; 259). Landsteiner and Popper had found a satisfactory model for experimental poliomyelitis. It was the first step in a very long saga which was to culminate many years later in the successful development of a vaccine and ultimate control of the disease (cf. Chapter 5).

In 1908, that was very much in the future. The facilities in the pathology laboratories in Vienna were modest, and Landsteiner realized that in spite of this promising beginning he had no hope of making rapid progress. He began a collaboration with Levaditi in Metchnikoff's department at the Institut Pasteur in Paris. The following year they could report successful filtration of the virus through a Berkefeld V filter (256; 257). Almost simultaneously, the same result was obtained by Flexner and Lewis at the Rockefeller Institute for Medical Research, who had enthusiastically entered the field at the first news of the successful monkey experiments (158). Within a very short time, Landsteiner discontinued his polio research altogether. It has been suggested that, being the archetype of the unselfish and altruistic scholar, he was content to leave the field to those whose facilities and financial resources made it likely that they would produce results sooner for the benefit of mankind. It is also possible that Landsteiner was anxious to devote himself more fully to his work on iso-antibodies. Virology's loss was the gain of immunology, and the study of experimental poliomyelitis passed to other hands.

5. Vaccination and vaccines

Because of the essential difference in nature and mode of replication between bacteria and viruses, viral chemotherapy (28) has so far made much less headway than its bacterial counterpart. Consequently, hopes of combating virus disease rest, now as in the past, mainly on the development of vaccines. Three anti-viral vaccines have elicited public and professional interest to an incomparable extent both during and after their development. Chronologically and methodologically, they comprise an impressive and instructive progression.

The earliest example, and the one to which we owe the term 'vaccination', was of course Jenner's cow pox vaccination. It will always remain a prize example of an eminently successful outcome to a purely empirical approach by a man who was above all an observer drawing intelligent conclusions, rather than a careful experimentalist. His was an age when the understanding of the aetiology of any infectious disease, let alone a virus disease, was fraught with uncertainty. Yet Edward Jenner, with little regard for the intricacies of painstaking experimental methods, relying on nothing so much as common sense and a good deal of luck, cut through all the problems he had not even thought of, and developed a prophylactic method based on what was to all extents and purposes an attenuated form of the virus. It was a remarkable achievement.

The subsequent stormy and long drawn-out history of vaccination was the inevitable outcome of the use of an empirical method for which there was no theoretical explanation at the time. In spite of his lack of knowledge of the facts behind his discovery, Jenner succeeded in laying the foundations of all later methods of prophylactic inoculation. There were to be many setbacks and reversals along the way. Within 12 months of the first mass vaccinations in the London Smallpox Hospital the identity of vaccine lymph had become obscured beyond the point of no return, perhaps by hybridization within host cells invaded simultaneously by cow pox and variola viruses. Vaccinia virus, morphologically and antigenically indistinguishable from the viruses of cow pox, variola major and variola minor, may elude an accurate definition, but it has all but conquered a centuries-old scourge of mankind. Jenner may not have been the most careful of experimentalists, or the most informed of theoreticians; but when we acknowledge our present dependence on prophylactic methods as our best weapon against most virus diseases we must also acknowledge that Edward Jenner showed the way with his modest pamphlet in 1798 (233). The development of a vaccine against rabies nearly a century later was a totally different undertaking both in conception and execution. It had to be, since both the diseases and the circumstances were so different. Smallpox, while not necessarily fatal, had been one of the major scourges of mankind for centuries; attempts to stem its tide had been made for nearly a century before Jenner, by inoculation of material which it was hoped would produce a mild form of the disease. Gatti (174; see p. 6) had perceptively and prophetically written of the need for a way of attenuating smallpox virus by passing it through a 'number of bodies', but had had no hope of putting his theory into practice.

Rabies, on the other hand, had been known for centuries to be inevitably fatal in man once the clinical disease had become manifest. Although the number of victims in any outbreak was trivial compared to the great epidemics of smallpox and bubonic plague, the terrifying clinical manifestations had given to rabies unparalleled publicity since the time of Aristotle. Louis Pasteur had had personal experience, at a tender age, of the despondency created by an outbreak in a rural community.[1] By 1880, a chance observation coupled with his own ingenuity had made possible the development of a vaccine against fowl cholera, based on the initially fortuitous attenuation of a virulent strain

---

[1] This and many other dramatic episodes are vividly described in the life of Pasteur published by his son-in-law, Vallery-Radot, in 1900 (455).

(347). This was followed by the less haphazard, more deliberately organized, onslaught on anthrax. The principle of attenuation had been established by accidental ageing of fowl cholera cultures; reasoning based on the comparative pathology of anthrax in birds and mammals led to the use of temperature-induced attenuation of the anthrax bacillus. Again, Pasteur was triumphantly successful (350).

By then, he was already deeply involved in experiments with rabies. In the cases of fowl cholera and anthrax, the aetiology of the diseases was well understood by the time vaccines were developed. This was far from being true of rabies. After four years of experiments, Pasteur could distinguish histopathologically between rabid and non-rabid brain material, but he was not able to give a rational explanation for his interpretation, nor could he claim to have isolated, let alone seen, a causative agent. It is a measure of his genius and his dogged determination that this did not deter him. Unable to isolate the mysterious pathogen, unable to grow it *in vitro*, he grew it, unseen but manifest, in its natural habitat: in the spinal cord of his laboratory rabbits, which a nineteenth-century tradition had established as a convenient host for experimental rabies.

G. G. Zinke in Jena had been the first to record, in 1804, the susceptibility of the rabbit to rabies, transmitted by inoculation of saliva from a rabid dog (495). John Hunter had suggested such an approach 10 years earlier, in a journal published in London (224). Zinke certainly knew of this work; whether it was his direct inspiration, he does not tell his readers. Together, Hunter's suggestions and Zinke's experiments constitute the first recorded, rational attempt to prove the transmission of rabies from dog to dog, and to other mammals including rabbits, and even to fowl.[1] In Paris shortly afterwards François Magendie showed, in a series of dramatic and dramatically presented experiments between 1810 and 1820 (303), that he could transmit rabies to dogs via the saliva from a human case. The involvement of Magendie, who was above all a neurophysiologist, in rabies research at this time coincided with a growing awareness of the neurotropic character of the pathogen. In the 1820s, Hertwig

[1] Hunter's suggestions and Zinke's experiments are compared and discussed in more detail in (475).

54

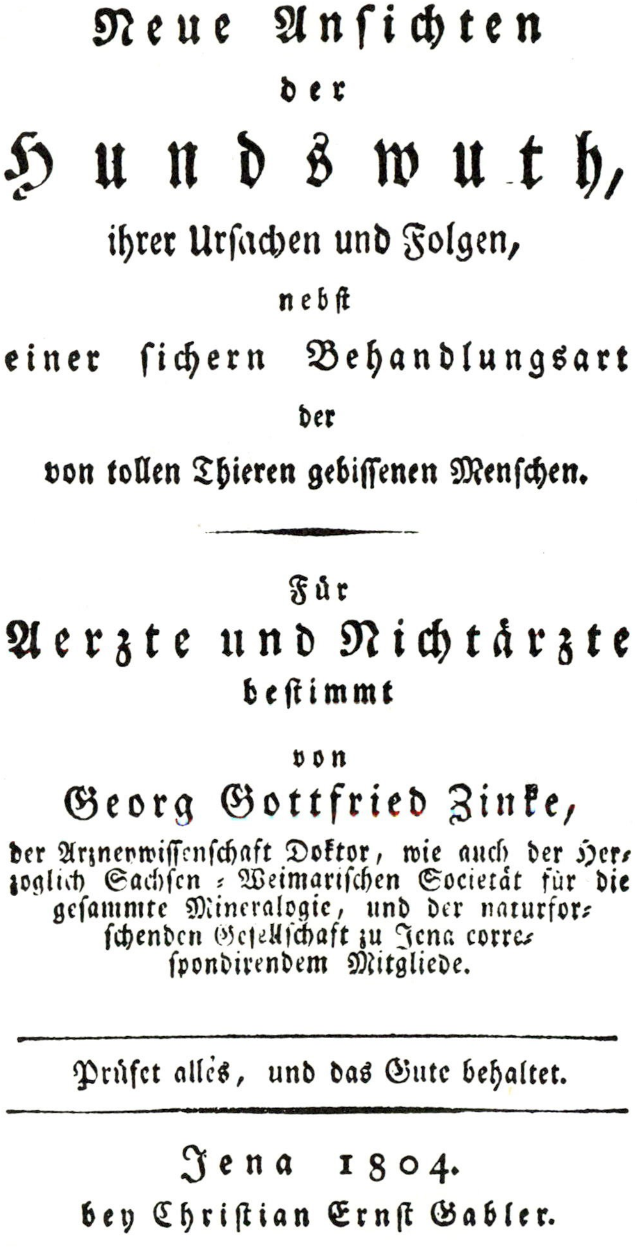

Neue Ansichten

der

Hundswuth,

ihrer Ursachen und Folgen,

nebst

einer sichern Behandlungsart

der

von tollen Thieren gebissenen Menschen.

Für

Aerzte und Nichtärzte
bestimmt

von

Georg Gottfried Zinke,

der Arznewissenschaft Doktor, wie auch der Herzoglich Sachsen = Weimarischen Societät für die gesammte Mineralogie, und der naturforschenden Gesellschaft zu Jena correspondirendem Mitgliede.

Prüfet alles, und das Gute behaltet.

Jena 1804.
bey Christian Ernst Gabler.

Fig. 14. Title page of Zinke's (1804) treatise on rabies (495). (Reproduced by courtesy of the Wellcome Trustees.)

55

in Berlin made a number of attempts, albeit unsuccessful ones, to transmit rabies by the implantation of nervous tissue from rabid animals into healthy ones (213).

Magendie and Hertwig chiefly used dogs (and a few cats) for their experiments. By the middle of the nineteenth century an English veterinary surgeon, William Youatt, regretted in print that he had neglected the opportunity to test rabbits as experimental animals in work on rabies. He clearly saw their potential value (492), but unfortunately only when his working life was over. Thus it was left to Galtier to reintroduce, finally and effectively, the rabbit into this type of experimentation in 1879. In careful and extensive experiments Galtier demonstrated the transmissibility of rabies from dog to rabbit, and from rabbit to rabbit in series, and pointed out the advantage of using rabbits with their predominantly paralytic and convulsive response rather than the 'furious rage'. Realizing the potential importance of his results, Galtier submitted a note to the Academy of Sciences in Paris (171) in addition to the full report he published in a veterinary journal (170). He then proceeded to experiment with the possible immunizing effects[1] of intravenous injection of rabid material (172). Among those who took notice of Galtier's work was Louis Pasteur.

Pasteur first established beyond doubt the neurotropic character of the virus. His next step was to achieve a reduced and predictable incubation period, by serial passage in rabbits. It eventually became constant at seven days (351). Pasteur had arrived at what he called a fixed virus, with a completely predictable incubation period. The problem then was to attenuate the fixed virus. Pasteur returned to attenuation by ageing; only this time he could not experiment on cultures. He had no alternative but to use the spinal cords of sacrificed rabbits. They were mounted in plugged glass containers with a small amount of caustic soda to act as a desiccating agent (Fig. 15). At the end of 14 days, the nervous tissue had become inactive. By using carefully graded material, beginning with suspensions of medulla which had become totally inactivated after 14 days, and

---

[1] Théodoridès has pointed out that the application in the case of rabies of the principle used in smallpox vaccination had been advocated in France 20 years earlier in theoretical considerations by LeCoeur in 1856, and by Vernois in 1863 (445).

working backwards through daily doses increasing in virulence until, on the fourteenth day, the inoculation contained fully virulent material from a freshly killed, rabid rabbit, Pasteur was able to render dogs completely refractive to experimental rabies. In September 1884, he wrote to the Emperor of Brazil, who took a great interest in his research: '. . . my studies on hydrophobia . . . are making good and uninterrupted progress. However, it will probably take another two years before I can bring them to a satisfactory conclusion and put them to practical use. . . But however often I can successfully repeat rabies prophylaxis in dogs, I fear my hand will tremble when I have to go on to man. . .'[1] His hand was forced long before the two years were up.

At the beginning of July 1885 young Joseph Meister, nine years old, was savaged by a rabid dog on his way to school in a small town in Alsace. The local physician advised his mother to take him at once to Paris, to see 'one who was not a physician, but who would be the best judge of what could be done in such a serious case'. Pasteur, faced with the small boy who was in great pain from multiple, serious bite wounds, was faced also with the most agonizing decision of his career. He had at his disposal a post-exposure vaccine which he knew to be efficacious with dogs; it had never been tested in humans. Yet there was no time to lose in speculation or ethical deliberation. Pasteur consulted Vulpian, the neurophysiologist who had been a member of the government commission formed to evaluate the results of anti-rabies vaccination, and who had become a trusted friend, and Dr Grancher, who worked in the laboratory. It was now more than 48 hours since the child had been bitten, and when the two men had examined Joseph Meister and seen the extent and severity of the wounds, they did not hesitate to support Pasteur in the decision to administer a course of post-exposure pro-phylactic inoculations immediately.

After two weeks of injections of increasing strength, Pasteur's anguish increasing with the increasing virulence of the ino-culum, the child's wounds were healing; he continued in perfect health. The night before the last, fully virulent, injection,

---

[1] The letter is reproduced in its entirety in Vallery-Radot's life of Pasteur (455).

Pasteur was unable to sleep, tormented by visions of what might conceivably happen to Joseph Meister. Nothing happened. Little Joseph survived, and so did Jean-Baptiste Jupille, the brave shepherd boy from the Jura who was treated by Pasteur three months later. These two initial successes established Pasteur in the eyes of the world as the saviour of those bitten by mad dogs. They came to his laboratories in large numbers, and from as far away as Russia. Inevitably, failures occurred, some of them only too predictable when too long an interval had elapsed between the bite and the arrival of the patient in Paris. Pasteur never turned anyone away, however hopeless the prognosis; his critics pounced on the failures as justification for denouncing Pasteur and his methods. However, majority opinion was favourable, and the Institut Pasteur was funded by public subscription in France and abroad. The institute was inaugurated on 14 November 1888; publication of its *Annales* had begun the previous year, before completion of the construction work.

Anti-rabies vaccine was far from perfect in the early days; 90 years later it is still far from perfect. There have been many setbacks and many problems which even the introduction of steadily improving tissue culture methods have failed to solve. But in view of the gravity of the prognosis and the disease there has been no doubt of the compelling need to use the vaccine on patients bitten by rabid animals.

Polio vaccine became an accepted reality 70 years after Pasteur first successfully practised rabies prophylaxis. The virus had been transmitted to monkeys and filtered in 1909. Why did it take so long? The reasons were manifold. In the case of rabies, vaccine is a last-resort, post-exposure, prophylactic measure. There was little to lose by treating Joseph Meister and Jean-Baptiste Jupille; untreated, they would almost certainly have developed the clinical disease and suffered the inevitable outcome. Poliomyelitis, like most other contagious virus diseases, was always unpredictable both in its epidemiology and in the severity of the frank clinical disease in individual cases. The use of any vaccine which left room for any doubt whatsoever would have been indefensible. Even with all the precautions taken, there have been several distressing incidents since the vaccines

59

Fig. 16. Pasteur in Copenhagen in 1884, when his attempts to develop a vaccine against rabies were moving towards a successful conclusion. (Reproduced by courtesy of the Musée Pasteur, Paris.)

have come into general use. But there were other reasons for the long delay.

Behind the initial excitement created by Landsteiner's discovery of the susceptibility of monkeys to experimental poliomyelitis (p. 50) lay the hopes of eventually controlling the disease by the development of a vaccine. In 1909, Pasteur's striking success in developing vaccines against a variety of infectious diseases of animals and man was fresh in everybody's mind. Hopes of equal success in the case of poliomyelitis were high, and estimates of the time required to attain such success were optimistic. In the spring of 1911, Simon Flexner was reported by the *New York Times* to have said: 'We have already discovered how to prevent the disease, and the achievement of a cure, I may conservatively say, is not now far distant. . .'[1] His talk of prevention presumably referred to the demonstration of antibodies in convalescent serum of monkeys and humans. Even so, and even allowing for journalistic licence, this was over-confidence on a large scale. When Flexner died in 1946, a vaccine against poliomyelitis was still far from being a reality.

During the years separating the development of rabies vaccine and the beginnings of work on poliomyelitis prophylaxis, a great many promising discoveries had been made in immunology which should be helpful to those striving to develop vaccines, but vital parts of the puzzle were still missing. In the case of poliomyelitis, it seemed almost the unfortunate situation of too many cooks spoiling the broth. The names of those associated with early and later efforts are legion; but according to J. R. Paul, who was in an ideal position to observe and comment (see note 1 below), the long delay was also the outcome of a strangely short-sighted policy of deliberate and total separation of those in a position to make clinical observations from those working with the experimental disease. The gulf between this attitude in the United States, where financial restrictions were less inhi-

---

[1] Quoted by J. R. Paul in *A history of poliomyelitis* (356), a comprehensive and riveting account, written in the last few years of his life by an uncompromisingly honest scholar who had himself been deeply involved in work on the disease. In 1937–38 he and J. D. Trask finally confirmed the Swedish results of 20 years before which Flexner had resisted for so long, establishing beyond doubt the presence of poliovirus in the intestinal tract. In the 1940s and 1950s Paul made contributions of importance for the development of vaccines against polio (see 219).

biting than elsewhere, and the more comprehensive approach by smaller isolated groups, above all in Sweden, was as deep as it was unfortunate.

Flexner remained convinced all his life that poliovirus was exclusively neurotropic, that it invaded its host via the respiratory system and that from there it went straight to the central nervous system. It was observed in a previous chapter that Landsteiner's accidental selection of monkeys was a fortunate one; it also had a negative side to it. The Rhesus monkey is highly susceptible to infection with poliomyelitis, but only via the respiratory system; it is all but insusceptible to infection via the alimentary canal. This supported Flexner in his rigidly held beliefs. With the weight of the prestige of the Rockefeller Institute behind him, he swayed majority opinion in this wrong direction for many years, and the voice of the Scandinavians, above all Carl Kling, was effectively drowned.

From the time of the early extensive Swedish epidemics, a great deal of important work had issued from Stockholm in particular. Carl Kling had first entered the field, with Pettersson and Wernstedt, at the time of the overwhelming Swedish outbreak of 1911. Unhappily, this epidemic yielded a large number of fatal cases, and the Swedish team was able to examine systematically a variety of anatomical sites. In a careful and extensive study, they recovered the virus not only from the expected locations in the oropharynx and the trachea, but also from the intestinal wall and intestinal contents. Kling and his group also tackled the question of healthy carriers, and isolated poliovirus from healthy members of the families of patients as well as from healthy members of the public without contact with frank cases (243).

In retrospect, it is easy to see that the Swedish findings were of paramount importance. At the time of publication, they elicited much less interest than they deserved, and were given no immediate chance to contribute to the development of a vaccine.

Another major stumbling block, and one which neither Flexner nor Landsteiner lived to see resolved, was the complexity of the family of polioviruses. Whereas cowpox gives protection against smallpox, and the viruses of vaccinia and smallpox are so similar serologically that we may never be able to unravel their taxonomic relationship, and whereas rabies virus appears to have remained astonishingly unchanged for 2000 years and to have no close relatives serologically, it was ultimately demonstrated that the family of polioviruses consists of three major strains, all serologically distinct. But the realization of this state of affairs, so fundamental to vaccine production, only began after 1930, and it was cleared up only after the painstaking work of the typing committee set up by the National Foundation for Infantile Paralysis in the United States in 1949. The achievements of this mammoth undertaking become all the more impressive when one remembers that the investigation ultimately included the typing of a total of 196 strains collected from near and far, and that all testing could be done only on monkeys. Although tissue culture was about to come into its own with the work of Enders, Weller and Robbins (cf. p. 76) it was perfected just too late to be of use in this major study.

The joint impact of the typing committee's data on serotypes, and the success of Enders and his young colleagues in growing poliovirus in cultures of non-nervous tissues (143), removed the final serious obstacles to vaccine production. Ever since 1909 the monkey had been the only animal host available for experimental poliomyelitis. Monkeys are expensive, and not in unlimited supply. The adoption of the rabbit as host for rabies virus greatly helped rabies research in the nineteenth century, and cleared the way for the development of a vaccine. Cultivation of poliovirus in cultures of human embryonic non-nervous tissue similarly facilitated the development of polio vaccine. The accompanying discovery of characteristic cytopathic changes, easily seen in the light microscope, and of pH changes of the medium, became enormously important in quantitative determinations, neutralization tests *in vitro* and several other areas which are important not only in the study of polioviruses, but also for virology in general.

Hence the work of Enders, Weller and Robbins became a watershed in the history of poliomyelitis, and made possible a rational approach to the question of vaccination. A highly organized attempt at developing a vaccine had already started with the typing programme launched by the National Foundation in New York. It continued in a no less organized manner, to large-scale field trials by 1954 with the formalin-inactivated Salk-type vaccine. If the anxieties of those responsible were no less than Pasteur's in 1885, responsibility was at least shared by a large number of individuals who in turn were backed by an impressive array of official bodies headed by the National Foundation for Infantile Paralysis and the United States Public Health Service. Even the distressing interlude in 1955, when certain lots of vaccine used in California turned out to contain live virus causing vaccine-associated cases of poliomyelitis, could not reverse the overall confidence. From 1962 onwards, the Sabin-type oral vaccines, using live virus attenuated through tissue-culture passage, began to take over. The principle was not new. Max Theiler had originally used it when he attenuated yellow fever virus for vaccine production by passage in mice (442); in the 1940s he showed that it was possible to immunize monkeys with a Lansing poliovirus strain which was no longer paralytogenic after passage through mice. Theiler did not continue his polio researches, but Hilary Koprowski adopted the idea by 1950, and even attempted to immunize a small group of volunteers with a strain which had been passaged through cotton rats (251). However, shortly afterwards Sabin could announce successful immunization of monkeys with virus modified by passage in monkey kidney tissue culture (389). Since 1962, Sabin-type oral vaccines have been used in the United States as well as in Britain and several European countries.

To a large extent, the three vaccines discussed above have made possible the control of those three virus diseases. On the other hand, problems persist, and the more facts are unearthed by current work in virology and immunology, the more apparent become the shortcomings and the hazards which will always be associated with large-scale immunization programmes, especially with live vaccines.

When Max Theiler demonstrated that he could modify poliovirus strains by passage through mice as he had done with yellow fever virus in order to develop a vaccine, it was not the only parallel between vaccine work on the two viruses. The formalin-inactivation practised in the preparation of the Salk-type vaccine had also been tried out in yellow fever work by Findlay and Hindle in 1929 (see 344), just before Theiler succeeded in transmitting the virus to mice. Yellow fever vaccine came into its own in large-scale vaccinations of troops going to tropical theatres of war during World War II. Early results were hardly encouraging; it was soon realized that serious outbreaks of serum hepatitis could be traced back to certain batches of vaccine (cf. Chapter 14, p. 171). If this was a setback for yellow fever vaccine, it contributed in no small way to our understanding of the difficult problem of viral hepatitis. The two types of viral hepatitis, infectious hepatitis or hepatitis A, and serum hepatitis or hepatitis B, have been increasingly in the searchlight ever since. Today, the respective viruses responsible have been seen in the electron microscope and the development of vaccines remains a desirable but not immediate possibility.

Enders, who had dropped out of the Harvard Medical School in order to pursue more general biology, returned to medicine with a vengeance when his work on tissue culture led straight to the development of vaccines against poliomyelitis. Nor was that his only contribution to control of viral diseases. With Peebles, he isolated measles virus in tissue culture in 1954 (144), and a successful vaccine was eventually developed and is now widely used. Again, the ethical problems were considerable. Post-vaccination encephalitis does occur, but much less frequently than encephalitis following natural measles; nor is there any evidence of increase in the very remote threat from sub-acute sclerosing panencephalitis.

Another important step was taken by Thomas Weller when, with Neva, he succeeded in isolating rubella virus in 1962 (473), and thus paved the way for the development of a vaccine. Rubella had then been known as a separate entity for a century and a half, described as distinct from scarlatina by Maton in 1815 (310), and as distinct from scarlatina and measles by Wagner in Germany in 1829 (see 311). The name rubella was suggested by Veale in 1866 (456). The mildest of eruptive fevers, it had been considered of little importance since, and the need for the development of a vaccine was not apparent until 1941 (see Chapter 12, p. 150).

In that year, Gregg in Australia discovered and described a

correlation between maternal rubella and a high proportion of congenital eye defects in infants (181). Gregg's results were amply confirmed and extended, and it soon became clear that the development of a vaccine against rubella deserved to be given high priority.

New problems will continue to arise as vaccines solve some of the older problems of controlling infectious disease.[1] Whatever the future problems of virology may turn out to be, the techniques of tissue culture should prove a valuable ally.

[1] The potential of interferon, first described by Isaacs and Lindenmann in 1957 (225), in the control of viral disease has not yet been adequately assessed, in spite of much competent work in this field. Very recently, encouraging results have been reported with the use of anti-viral antibodies in virus-induced cat sarcomas (103).

6. Development of cell and tissue culture

In the last decades of the nineteenth century, hopes were high among practitioners of medical and veterinary sciences that the infectious diseases might some day, perhaps in the not too distant future, be eradicated. Koch's postulates,[1] his culture methods and his photomicrography and other methods of identifying micro-organisms, together with Pasteur's development of vaccines, the early work on immunity, the identification of bacterial toxins and anti-toxins – all the exciting discoveries and inventions of the closing decades of the nineteenth century – combined to make this appear a not too impossible dream.

Then came the discovery of the filterable viruses, and the dream began to recede. It has been receding ever since, until increasing knowledge and even more sophisticated techniques have now brought us to a crossroads where new and no less serious anxieties plague those who are most thoroughly familiar with the subject. The problem of the accidental creation, in or out of the laboratory, of previously unknown, more deadly and invincible species of micro-organisms and viruses is as unpleasant as it is difficult to plan against.

But at the turn of the century such alarming possibilities had not yet occurred to anybody, and the problems were more immediate. The most immediate problem encountered by the discoverers of the first filterable viruses was the failure of these

---

[1] The eponymous criteria to be met in order to establish a particular micro-organism as the specific aetiological agent of a particular disease were: (1) the pathogen must be found in all cases of the disease; (2) it must be possible to isolate and maintain it in pure culture; (3) healthy, susceptible animals must develop the specific disease upon inoculation of the pure culture; (4) it must be possible to re-isolate the pathogen from such animals and again grow it in pure culture. Formulated by Koch in works on tuberculosis in the early 1880s, the principles had been laid down in theoretical terms by Koch's teacher at Göttingen, Jacob Henle, 40 years earlier (see 149).

disease agents to grow on any conventional artificial media, either liquid or solid. In the early days, a majority of those working in this field considered it merely a temporary difficulty, to be solved eventually by the development of other, more suitable media. Beijerinck, with his keen perception, thought otherwise. He made it clear from his very first paper on the subject that in his opinion the agent of tobacco mosaic disease did not only show a predilection for young, actively growing tissues, but that in view of its special nature it would never in fact be found to grow outside the living cell at all.

It was a view which became accepted slowly and gradually, and which could be ultimately justified and provided with a theoretical explanation only after the essential nature of the filterable viruses had been revealed. The tool which eventually made possible the rational cultivation of viruses for a variety of purposes has become known as tissue culture. Its origins are to be found in the nineteenth century, above all in the work of Wilhelm Roux (387), and in the briefer and less publicized investigations of Laurent Marie Chabry (see 81). Roux's contribution to the study of embryonic development represented a major departure from what had until 1880 passed for embryology, and married this discipline to the emerging experimental physiological chemistry rather than to the post-Darwinian speculative morphology, which had sought merely to use morphological observation of embryonic changes to explain facets of evolutionary history. Chabry died in 1895, and left no school; by then he had in any case joined the Institut Pasteur and was working on tuberculosis. Wilhelm Roux, on the other hand, left no stone unturned in his efforts to publicize the new experimental discipline he had created, and which he called *Entwicklungsmechanik*. True to his mould of a great nineteenth-century German natural scientist, he founded an *Archiv für Entwicklungsmechanik der Organismen*, and was provided with his own Institut für Entwicklungsmechanik.

Not surprisingly, his thoughts and experimentation influenced a whole generation of physiologists and anatomists, not only in Europe, but also in America. In 1894 the Marine Biological Laboratory at Woods Hole, Massachusetts, inaugurated a celebrated series of Friday evening lectures with a translation of Roux's 'manifesto'[1] from the initial issue of the *Archiv*. The director of the laboratory impressed upon his listeners the need for a new approach in 'biological physiology' along the lines of Roux's experiments on frog embryos. In the same year a young man presented his doctoral thesis at the Johns Hopkins in Baltimore; he went on, on leave from an instructorship at Johns Hopkins, to qualify as M.D. in Bonn, and returned as professor of anatomy to his *Alma Mater* (see 2). He was to stay there for the rest of his active academic career, and it was here that he made methodological history in 1907 with the kind of experimentation which could be traced back to the influence of Roux. His paper published in that year in the *Proceedings of the Society for Experimental Biology and Medicine* had the modest title of 'Observations on the living developing nerve fiber' (198). Its contents, with its successors over the next three years, were to revolutionize the methodological approach to a whole range of problems. By the time Ross G. Harrison read the final paper of the series, on 'The cultivation of tissues in extraneous media as a method of morphogenetic study' to the American Association of Anatomists in December 1911 (199), it was the opening address of a symposium billed under the general title of 'Tissue culture'. By then, Harrison's 'hanging drop' method had already become a classic technique.

Alexis Carrel had lost no time in sending his assistant, M. T. Burrows, to Harrison to learn the new technique, which until the 1920s remained the coverslip method originally designed by Harrison. But where Harrison had developed the technique in order to study tissue growth at the cellular level, Carrel's interest in tissue culture was very differently motivated. His concern was to find a method which would allow the survival of whole organs *in vitro*,[2] and by 1923 his flask culture (69) grew naturally out

---

[1] Wilhelm Roux defined his '*Entwicklungsmechanik*' as the science of the reasons for organic shapes and conditions, including the causes of their formation, maintenance and repair. He warned against invoking metaphysical reasons for processes hard to understand and interpret, reminding his fellow scholars that,' *Incidit in scyllam, qui vult vitare charybdim*' (387).

[2] Tissue culture was for Alexis Carrel a step along the way from his work on vascular suture and transplantation of blood vessels and organs in animals to the later much publicized collaboration with Charles Lindbergh on an artificial heart pump. For details of Carrel's sometimes controversial, always colourful, career see, for example, the account by Edwards and Edwards (135).

particular cell of origin can not be distinguished. The figure represents two stages of the same fiber sketched at an interval of twenty-five minutes, during which time the fibre has lengthened twenty microns. The case shown in Fig. 22 is a much larger fiber, about 3 microns in

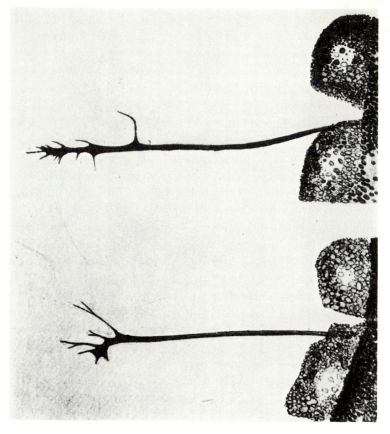

Fɪɢ. 22.—Two views of the same nerve fiber, taken fifty minutes apart. Preparation similar to that shown in Figs. 20 and 21.

diameter, with much more protoplasm at the end. The movements of this fiber were extremely active, and the change of form with accompanying lengthening is well shown by comparing the two sketches, which were made fifty minutes apart.

Fig. 17. R. G. Harrison's (1908) figure of a growing nerve fibre in tissue culture, from *Anatomical Record*. (Reproduced by courtesy of the Royal Society of Medicine.)

70

of the need to accommodate larger fragments of tissue. Another of his preoccupations was with malignant growths, and the two interests came together in 1926, when he could show that a filtered extract of Rous' chicken sarcoma would proliferate in a medium consisting mainly of chicken serum and Tyrode solution only if it also contained fragments of fresh tissue (71). Two years earlier he had found that leucocytes would also support the proliferation of Rous' sarcoma virus (70), a conclusion which brings to mind Landsteiner's suggestion that Marchoux's results with fowl plague virus in 1908 were the outcome of the presence of whole blood cells rather than any substance which could be chemically defined (255; see p. 42).

The problem of *in vitro* culture of viruses has had a profitable interrelationship with the development of tissue culture methods. In a sense, they grew up together. A younger colleague of Ross Harrison's in the anatomy department at Johns Hopkins, Warren H. Lewis, married in 1910 Margaret Reed, a biologist who had recently returned from a stay in Berlin. There she had been impressed by the work of Rhoda Erdmann, who at that time was attempting to cultivate amoeba on a nutrient agar made up with a physiologically balanced salt solution. A few years later (1916) Rhoda Erdmann showed that it was possible to keep fowl plague virus alive for about a week in a tissue culture of red bone marrow (146). But Margaret Reed was inspired by Rhoda Erdmann's early experiments to explant a small fragment of guinea pig bone marrow into the same medium (see 201). After a few days in an incubator the bone marrow cells could be seen to form a membranous growth on the surface of the agar, with some nuclei undergoing mitosis, i.e. the cells were not only surviving, but actively multiplying. This may have been the first *in vitro* culture of mammalian cells, pre-dating Carrel, and it turned her attention, and that of Warren H. Lewis after their marriage, to the growing of cells, and discrete tissue fragments, in culture. They made many contributions to the development of culture media, but their hopes of achieving culture media which were well defined in chemical terms were premature and could not succeed at a time when little was known of vitamins, trace elements and other growth regulators.

71

Cell and tissue cultures did not at first yield any spectacular advances in the study of viruses. Rather, tissue culture *per se* and tissue culture as applied to the growing and studying of viruses developed together, influencing each other along the way, and both benefiting from the relationship. Just before and during World War I the first attempts were made to show that viruses could remain active in cells surviving in culture. Techniques were still relatively primitive and results were few and far between. Apart from the early experiments with fowl plague virus by Marchoux and by Erdmann (306; 146), most attempts at cultivation were concerned with the viruses of vaccinia, poliomyelitis and rabies. Levaditi, following his collaboration with Landsteiner when they first filtered poliovirus, proceeded to attempt to cultivate the virus in spinal ganglion cells (267), as he also tried to do with the virus of rabies (268). It is an interesting reflection upon the preoccupations at the Institut Pasteur in its initial decades that the first paper in which Levaditi and Landsteiner announced the filtration of poliovirus (256) had a gratuitous *arrière-pensée* which has not wholly stood the test of time: 'There is a striking analogy between the viruses of rabies and acute poliomyelitis'. Levaditi's two papers describing his early forays into the tissue culture of viruses appeared in 1913. The same year, Noguchi, who later with such tragic results concentrated on yellow fever research (cf. p. 145), published suggestions for what he felt confident was a method of cultivating rabies virus using a technique he had developed for growing certain spirochaetes (330). Persistence of the virus through a few passages rather than replication was all that could be hoped for in these early attempts.

Fifteen years later, little progress had been made. In his chapter on 'Tissue cultures in the study of viruses' in the classic text on filterable viruses edited by Thomas Rivers in 1928 (370), Alexis Carrel wrote: 'Through the rudimentary techniques of the early days of tissue culture...it was demonstrated that tissues kept *in vitro* can be utilized in the investigation of the properties of viruses. Although fourteen years have elapsed since this work was undertaken, the method of tissue culture has not greatly increased our knowledge of the heterogeneous group of filterable pathogenic principles...' And Carrel con-

72

tinued: 'This must be partly attributed to the fact that pathologists are far from having mastered the techniques for the cultivation of tissues. Most of them have used the comparatively crude procedure which was derived immediately from the experiments of Harrison...' In fact, Carrel himself had a few years earlier developed a method of growing in flasks fragments of tissue larger than those which could be accommodated by Harrison's coverslip technique. His review was written at a time when his flask culture technique was beginning to show results in the study of viruses.

Harrison had grown his first tissues in clotted frog lymph; Burrows and Carrel substituted at first fowl plasma and then a mixture of serum and embryo cells (66). From the early days, physiologically balanced salt solutions were frequently used to dilute the natural media. By the time Carrel introduced flask culture, he used a mixture of solid (fibrin coagulum) and fluid medium for the cultures, and diluted the plasma and tissue extracts used with Tyrode solution (69). Tyrode's modification (453) of the physiological salt solution originally used by Sidney Ringer in studies on the effects of crystalloids and colloids on the contractions of frog's heart[1] and skeletal muscle (368) has been extensively used as a diluting fluid for the natural, and, much later, the chemically defined media of tissue culture.

The introduction of flask culture gave new impetus to attempts to grow viruses in tissue culture. A number of early attempts had been made with the virus of vaccinia. Following Levaditi's work on the viruses of poliomyelitis and rabies in 1913, Edna Harde in New York found some proliferation of vaccinia virus in hanging drop cultures of rabbit cornea (197), and Parker and Nye could report positive results with the same virus in cultures of rabbit testis in rabbit plasma and Ringer solution (345). In 1927, Carrel and Rivers applied the flask culture technique to vaccinia virus, and waxed enthusiastic about their results, to the extent of speculating that possibly '...one finely pulped chicken embryo might be capable of producing as much vaccine virus as a calf...' (72).

[1] Ringer's saline solution of the 1880s (368) was first modified by F. S. Locke between 1895 and 1901 (274). Locke raised the salt content to obtain an osmotic pressure more in keeping with the body fluids of higher animals. In 1910, Tyrode improved the Ringer–Locke solution by the addition of sodium and magnesium phosphates (453).

The following year the Maitlands in Manchester used finely minced hen's kidney in hen's serum and Tyrode solution to obtain 25×10^6 times proliferation of vaccinia virus, even, as they put it, 'without tissue culture', since they could detect no growth of the kidney tissue used, only slow disintegration and eventual autolysis (304; cf. p. 144). In 1929, C. H. Andrewes applied the technique to the so-called virus III (10), the latent herpesvirus of rabbits which was discovered in 1923 (373), when it temporarily disguised the results of the attempts by Rivers and Tillett to transmit varicella to rabbits. In Germany, Hallauer used tissue culture in a study of fowl plague virus which yielded important results. Using pulped chicken embryos, he found that virus growth was sustained only in the presence of living tissues, and that the rate of virus replication depended on the degree of proliferation of the tissue. He also observed characteristic cytopathic effects, and suggested that eventually it might be possible to develop a method of virus titration in tissue culture to replace the more costly and cumbersome animal experiments (195).

Hallauer's paper was published in 1931, and that proved to be the year when, as a method, tissue culture of viruses was temporarily overtaken by events. Woodruff and Goodpasture showed that instead of relying on pulped chicken embryo tissue in culture, it was possible to grow certain viruses to great advantage on the sheets of uniform cells of the whole developing chicken embryo inside the fertilized egg (488). It had been tried before, but the results had been sporadic and desultory and had led nowhere. Monckton Copeman may or may not have grown vaccinia virus in eggs in 1898, but he was unaware of the nature of the agent he was experimenting with, and his results were unimportant (cf. p. 39). Centanni certainly grew fowl plague virus in embryonated eggs when he first showed the filterability of this pathogen at the turn of the century; he even observed characteristic deformities in two infected embryos, and thus came close to the theoretical background to modern methods of virus titration (see pp. 39–41). Murphy and Rous in 1911 found fowl sarcoma virus to proliferate in eggs (384).

In 1920, Jouan and Staub reported results of incomplete work on fowl plague which had been interrupted by World War I (239). They were primarily interested in the possibility of devel-

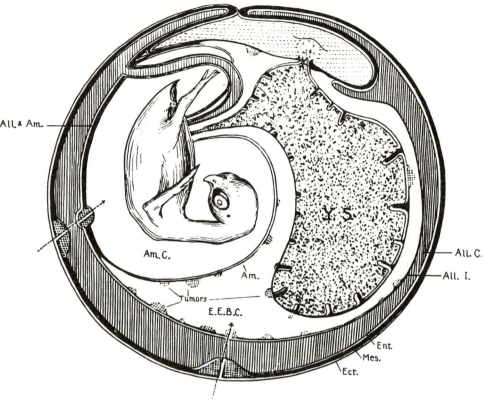

Labels on figure:
All & Am.
Am. C.
Am.
Tumors
E.E.B.C.
Y.S.
All. C.
All. I.
Ent.
Mes.
Ect.

Fig. 18. Murphy and Rous in 1912 (321) adapted an illustration from F. R. Lillie (1908), *The development of the chick*, Chicago, to demonstrate their method of inoculating the embryonated egg and its membranes. Later textbook versions have become increasingly diagrammatic.

oping a vaccine; while performing a post-mortem examination on an infected hen, they extracted a fully developed egg. Preserving the egg yolk in a test tube they found it to be virulent and to remain virulent for a period of weeks. This led them to inoculate embryonated eggs with the virus; their results were inconclusive and they were still struggling to find the optimum stage of development for the embryonated egg to be inoculated when they appear to have abandoned the project.

After these early attempts at egg culture involving the viruses of vaccinia and of fowl plague, it was in the end work with the virus of fowl pox which led to the successful development of

75

the method of growing viruses on the chorio-allantois of the embryonated egg. The egg can be seen as a particularly cheap and convenient experimental animal; by a stretch of imagination (and definition) it can perhaps also be seen as a very sophisticated kind of tissue culture, carrying its own medium, by the same token that W. Roux's frog embryo experiments are often seen as the beginnings of tissue culture.

However we choose to see it, it has become an enormously successful method, and found many uses since it was first described by Woodruff and Goodpasture in 1931. But there can be no doubt that its instant and resounding success deflected attention from the growth of viruses in other tissues at a crucial stage in its development.

Nevertheless, techniques were still being improved and adapted for specific purposes. In 1933, George Gey, who had served his apprenticeship in tissue culture with the Lewises at Johns Hopkins, introduced a modification which has become known as the 'roller-tube' method,[1] in which optimum conditions for oxygenation of the fluid are ensured by slow rotation of the tubes in a drum (176).

Tissue culture methods in virus research came finally and triumphantly into their own with the work of Enders, Weller and Robbins on the Lansing strain of poliomyelitis (143). This not only showed the way to the development of polio vaccines, it also served to establish a number of important facts which were to have immense impact on the study of both this and other viruses. Among their observations of particular general importance were the cytopathic effect of the virus on the cells in culture, and changes induced in the pH values of the culture fluid.

The principle of one infective particle–one lesion had long been a desirable but not realizable possibility in the field of animal virus studies. Guérin had attempted something of this nature with vaccinia virus on shaved rabbit skin as early as 1905 (187). That bacteriophages could be titrated by counting the plaques formed in lawns of bacterial cells became evident with

---

[1] Among Gey's other notable contributions was the introduction of the HeLa cell line (see 238) and the observation of discrete lesions produced in rat sarcoma cells by Eastern equine encephalitis virus (cf. p. 77).

the observations of d'Herelle (106; cf. p. 90); after 1931, Good-pasture's work showed that the pocks formed on the chorio-allantois of the developing chick embryo could be used for quantitative measurements of certain viruses (488).

But for many years 'research on the growth characteristics and genetic properties of animal viruses has stood greatly in need of improved quantitative techniques, such as those used in the related field of bacteriophage studies'. With this paragraph Dulbecco finally introduced such a technique in 1952 (126). The previous year, Gey and Bang, in the course of a study on applications of tissue culture, had noticed that discrete lesions were produced in sheets of rat sarcoma cells by eastern equine encephalitis virus inoculated at high dilution (177). Now Dulbecco presented a technique for 'production of plaques in monolayer tissue cultures by single particles of an animal virus';[1] he also showed that plaques of the two viruses he used for this initial study were mutually indistinguishable. An improved version of the plaque assay method was later described by P. D. Cooper (91).

With this technique, and with the work on cytopathic effects which came to fruition at about the same time, tissue culture and virus research together began to fulfil the promises of their joint usefulness which had been suspended for so long.

---

[1] Dulbecco later described how contemplation of his newly established lawn helped him formulate his ideas at two consecutive stages in the development of the plaque titration technique; in particular, colour changes in winter prompted his realization of the necessity for developing suitable staining methods for the plaques (128).

7. Early theories concerning the nature of viruses

When World War I broke out, the pursuit of the esoteric and loosely characterized filterable viruses temporarily gave way to the more pressing medical problems of dealing with wound infections and combating the outbreaks of infectious diseases inevitably caused by the conditions of major wars. But it would be wrong to suggest that virus research was interrupted at a crucial stage. The facts and the theories which had become established before 1914 remained the basis for work in this field for many years to come.

Beijerinck's theories have been discussed in a previous chapter; they were so far ahead of their time that they were afforded little more than curiosity value by most pathologists. For obvious reasons, his views received more attention from plant pathologists; but if some of them provisionally accepted the idea of the infective particle as a chemical molecule, they could conceive of it only in terms of an enzyme, perhaps accumulated to pathological levels under conditions of external stress (cf. Chapter 10). Most clinical pathologists were inclined to accept Loeffler and Frosch's concept of very small microbes, differing in size only from their more conventional counterparts. However, the failure of the filterable viruses to grow on artificial media increasingly worried a number of authors.

For many years, the filterable viruses as a group were distinguished only by three common negative characteristics: they were not retained by bacteriological filters; they could not be seen in the light microscope; and they could not be grown on artificial media.

To a generation of pathologists who were still excited by the spectacular successes ensuing from work with Koch's postulates and Koch's media, the latter characteristic was the most difficult to reconcile with their new-found creed. Many continued to believe that determined experimentation would eventually yield suitable media. But perceptive souls were beginning to suspect, with Beijerinck, that perhaps this particular circumstance was a central key to the concept of filterable viruses, and that filterability was merely a coincidental characteristic.[1] As early as 1908, M'Fadyean of the Royal Veterinary College had stated that 'Another characteristic common to all the ultra-visible organisms is that they appear to be obligatory parasites' (300). The previous year von Prowazek, defining the group he named chlamydozoa, wrote of their 'very specific, unusually and extensively adapted parasitism' (464).

As the number of agents recognized as filterable viruses grew, the need for some kind of classification based on common biological factors rather than mere size was increasingly felt. At a very early stage Borrel, primarily concerned with tumours (cf. Chapter 13) rather than filterable viruses, did make a departure in favour of classing together those viruses causing exanthematous symptoms, or what he called 'infectious epithelioses', i.e. vaccinia, variola, sheep pox, fowl pox, foot-and-mouth disease, and rinderpest (53). This was an attempt to use pathogenic effects in classification and to stress the affinity of certain viruses for certain types of host tissue, and as the first of many it became a model for much later work.

Von Prowazek's chlamydozoa, on the other hand, was a more heterogeneous group with regard to clinical manifestations; its common features were to be found by histological examination of affected tissues. Von Prowazek was originally a botanist whose flair for microscopy led him to histological work on tropical diseases. When he became interested in the filterable viruses, through work on trachoma, his microscopist's mind naturally gravitated to those agents which induced visible changes, i.e. inclusion bodies, in the host tissues at the cellular level, in

---

[1] It was eventually recognized that modification of the postulates was necessary in the case of diseases caused by viruses. In 1937 (372), T. M. Rivers formulated such modifications, giving examples of the wrong conclusions sometimes reached through blind adherence to these principles; he also pointed out that Koch himself had realized at an early date that there might be cases when it would prove impossible to meet all the conditions. A. S. Evans has recently summed up the fate of the postulates over the years (149).

particular vaccinia, variola, rabies, fowl plague, fowl pox, trachoma, etc. Staining inclusion bodies after Giemsa, he found
them to consist of two distinct components, namely tiny, well-
defined, reddish granules, surrounded by an amorphous blue
substance stained blue. Their structure appeared to be reminiscent of certain protozoa, and von Prowazek called them
chlamydozoa, or hooded or mantled animals (464). From the
beginning, it seems to have been a contradiction in terms, for
von Prowazek always assumed that the amorphous blue
substance was a reaction product of the cell formed in response
to the invasion by the granules for which he adopted the term
elementary bodies (first used by Chauveau in studies on vaccinia
long before the discovery of filterable viruses (cf. p. 7); von
Prowazek's own recognition of this limitation to his concept
probably accounts for his ready acceptance soon afterwards of
Lipschütz's modification which used the term strongyloplasma
for the same group to indicate that the granules only were
concerned in pathogenesis (271).

The terms chlamydozoa and strongyloplasma went into contemporary textbooks but otherwise contributed little to the
formation of the virus concept. Von Prowazek might have
contributed more had he survived World War I; but he died
in a German field hospital in 1915, at the age of 39, of the
typhoid fever whose aetiology he was trying to determine (see
193). The following year Roche-Lima isolated the agent and
called it *Rickettsia prowazeki* (376), preserving for posterity the
names of von Prowazek and of H. T. Ricketts who had similarly
succumbed to the infection he was studying five years earlier
(see 132).

It was the absence of visible biological manifestations which
so puzzled those studying the behaviour of viruses. Von Prowazek and Lipschütz thought at one point that they had
observed binary fission in the mantled animals, but their evidence was unconvincing and no one else could report similar
observations.

Nevertheless, the absence of alternative explanations made
for a strong temptation to identify as protozoa those viruses
which formed inclusion bodies. Negri died convinced that the
bodies named after him should be identified as a protozoal agent

of rabies (325). Attempts were even made to classify as protozoa certain structures thought to be associated with tobacco mosaic disease (cf. Chapter 10), and as late as 1924–26 Levaditi and others were busy inventing names and classifications for the 'protozoa' of rabies and of what they considered a related pathogen of an acute encephalitis in rabbits (269). However, while the inclusion bodies continued to cause confusion in the concept of the pathogens inducing this kind of response in infected tissues, discussion and conjecture were pursued in other areas and on different levels throughout this period.

From the very beginnings at the turn of the century, Beijerinck's bold pronouncements had set the scene for a fruitless but intensely pursued argument which was to rage for more than a quarter of a century. It is an argument which cannot yet be said to be entirely resolved, but which by now is acknowledged to be pointless rather than fruitless; or perhaps 'beside the point' would be more accurate. In its briefest and most basic form the question was: 'Viruses – dead or alive?'; a question to which there is no cut and dried answer, just as there is no easy solution to the problem of the origin of viruses. The more we know, the more hard facts we accumulate concerning viruses, the more we become obliged to qualify and modify any unequivocal definitions we may have been tempted to make.[1]

When the mysterious filterable and invisible viruses came to the notice of the world of general biology and pathology at the turn of the century, this aspect of the problems surrounding them soon became part of a wider pattern of discussion sparked off by new knowledge yielded by new approaches in physiology and biochemistry. Physiologists were beginning to cooperate with chemists and physicists in attempts to unravel the secrets of life at the cellular level. In these beginnings can be seen the seeds of molecular biology; it is no accident that the filterable viruses soon were included as an integral part of the discussion. Molecular biology and virology were to develop along parallel paths until the dividing lines began to blur with some of the

---

[1] Sir Macfarlane Burnet has repeatedly expounded his reasons for viewing viruses as organisms (62), and warned of the dangers inherent in a purely theoretical approach, with no regard for clinical or general biological implications, in work on laboratory strains of bacteria and bacteriophages (63).

discoveries in the areas of bacteriophage and tobacco mosaic virus in the 1940s and 1950s.

The last quarter of the nineteenth century had seen a great increase in the volume of knowledge of cell constituents. This led to speculation concerning the powers of survival of individual structures. Giglio-Tos of Turin, now largely forgotten, defined 'biomolecules' as the smallest quantities of living matter able to exhibit some of the chemical phenomena of life,[1] such as respiratory exchange, assimilation and starch formation – surely a forerunner of the lysosome concept. But one basic question must be answered before considering such detail. It was the question of the minimum volume of a living entity possessing all the molecules necessary to sustain the complex processes of life.

It was the minute dimensions of the filterable viruses coupled with their obvious infectious qualities which made them interesting in this context. Beijerinck had offered one explanation, and it came very close to the modern concept. He labelled the agent of tobacco mosaic disease a '*contagium vivum fluidum*', and maintained that it could not be cellular but must be able to exert its influence, and even to replicate, in solution (36). This led him to the even bolder hypothesis of a pathogenic molecule which somehow became incorporated in the host cell and borrowed its existing metabolic and replicating mechanisms for its own purposes. It is not surprising that contemporary scientists found it well-nigh impossible to accept such a concept in 1899. But within five years of the discovery of filterable viruses two papers appeared whose authors used a mathematical approach to consider the viability of organisms of such small dimensions and to try to place them in a general pattern of physiological evolution.

The first was delivered as an address to the physiology section of the British Association at its 1901 meeting by its president, John McKendrick (301), for 30 years professor of physiology at Glasgow University and author of several textbooks of physiology. The second was first published two years later, as a

---

[1] Giglio-Tos' concepts – he envisaged a whole structure based on biomolecules, which formed larger aggregates, biomones, which in turn were built into biomonads or bioplasm (protoplasm) – and their implications were discussed at length by McKendrick (301).

theoretical examination of possible lower limits to the size of living organisms (147), by Leo Errera, a Belgian botanist who made good use of mathematics and physical chemistry in theoretical observations on biological problems. Both McKendrick and Errera were familiar with the earlier speculations by Clerk Maxwell, first holder of the chair of experimental physics at Cambridge, and as such responsible for much of the early planning of the Cavendish Laboratory before his untimely death at the age of 47 in 1879 (150). Clerk Maxwell had been unable to reconcile his calculations of molecular dimensions with prevailing theories of heredity and the degree of differentiation ultimately resulting from the growth of initially microscopic germs of plants and animals into highly organized organisms.

Unaware of each other's papers, and both able to improve on Clerk Maxwell's calculations in the light of later developments, McKendrick and Errera arrived at very similar results in their efforts to determine the number of protein molecules contained in the smallest microscopically visible organisms. McKendrick was concerned with the general implications of a molecular concept of living tissues. He concluded that the smallest organisms seen under the microscope 'may contain as many as 1,250 molecules of such a substance as a proteid [sic]'. McKendrick based his calculations partly on arbitrary values which he adopted from the physicists of his day, i.e. '... Take the average diameter of a molecule as the millionth of a millimetre...' and partly on more realistic values representing the limits of resolution of contemporary microscopes.

To McKendrick, viruses came as an after-thought for which he relied on information provided by his friend James Ritchie. Writing from the Pathological Institute at Oxford, Dr Ritchie mentioned only three 'organisms too small to be seen by the microscope'; they were the pathogens of foot-and-mouth disease, pleuropneumonia and African horse sickness.[1] The Oxford pathologist either did not know of, or did not care to include, Beijerinck's postulates concerning tobacco mosaic virus. In view of McKendrick's conclusion, it would have been interesting to witness a confrontation between his emergent

---

[1] Dr Ritchie's 'Notes on Organisms too small to be seen by the Microscope' and his accompanying letter are quoted verbatim by McKendrick.

biophysics and Beijerinck's corresponding biochemistry. McKendrick wrote:

I am of opinion, therefore, that it is quite justifiable to assume that vitality may be associated with such small particles, and that we have by no means reached what may be called the vital unit when we examine either the most minute cell or even the smallest particle of protoplasm that can be seen. This supposition may ultimately be of service in the framing of a theory of vital action. (301)

Errera's essay was more directly derived from the discovery of filterable viruses. He had a long-standing interest in the application of the laws of physics and mathematics in cytology, especially with regard to the lower limits of size in organisms, and was the author of Errera's law. According to this rule the cell wall, at the moment of formation, has the qualities of a liquid film assuming the shape encompassing the smallest possible area.[1]

Errera chose as his title 'On the lower limits to the size of organisms', and his first paragraph established the newly discovered filterable viruses, including that of tobacco mosaic disease, as the centre of his argument. He went on to ask, in the botanical terms which came naturally to him: 'Is it conceivable that living organisms might exist which are smaller than bacteria by the same factor as that by which bacteria are smaller than human beings, let alone such magnificent species of the plant kingdom as the Giant Sequoias of California or certain Australian Eucalyptus trees?' (147).

Using calculations similar to McKendrick's, Errera was obliged to reply in the negative. In view of the discontinuity of an organized cellular structure, granted that living matter consists of small, but not infinitesimally small, molecules, Errera concluded that there must be a lower limit below which organisms cannot exist. However undifferentiated, they are still aggregates of complex molecules and must be of a finite size. On the basis of comparative calculations he found even organisms a few hundred times smaller than the smallest recorded bacteria to be inconceivable, and concluded that probably even

[1] Errera's versatility in life, work and philosophy is discussed in a series of essays in a memorial volume published by the Université Libre in Brussels in 1958 (148).

the 'invisible microbes' were not very much smaller than the smallest known micro-organisms.

The considerations of McKendrick and of Errera have long since been overtaken by events. They show very clearly the limits imposed on even inspired thinking of any period by the weak links in the sum total of contemporary scientific knowledge. On the credit side, they also show how very rapidly the filterable viruses became an accepted part of the theoretical background on which advances in biology must be built. Both Errera and McKendrick were well aware of the limitations to their work, and McKendrick wrote: '. . . physiology shares with the other sciences the possession of problems that, if I may use a paradox, seem to be more insoluble the nearer we approach their solution'.

The next step towards a solution of the mystery of the filterable viruses was taken in 1915. It admirably fitted McKendrick's paradox, and for many years served only to deepen the mystery. It was the discovery of bacteriophage.

8. The bacterial viruses
I

Bacteria are defined as unicellular organisms belonging to the plant kingdom. There is no obvious relationship between bacteria and other organisms among the lower plants; certain features suggest that they may be remotely related to blue-green algae. However, for a number of reasons, the viruses attacking bacteria are never included casually with other plant viruses. The reasons are both historical and, to an almost overwhelming extent, factual. Whereas viruses attacking blue-green algae have only recently come to light,[1] and have so far created no great excitement, bacterial viruses have been known, and extensively studied, for more than 60 years.

The history of the study of bacterial viruses falls into two separate parts. The early history spanned two decades and volumes of frequently sharply worded, academic argument from a few protagonists. The protagonists each had a favourite name for the phenomenon they were describing, and it was variously referred to as 'the lytic phenomenon', 'the Twort–d'Herelle phenomenon', 'bacteriophage' and 'transmissible autolysis'. The second phase began in the 1930s, when links were forged between such disparate disciplines as physics and biology, links which, through the concerted efforts of many astute scientists from all over the world, were to lead to an understanding of the nature of bacterial and other viruses, and eventually to the cracking of the genetic code. By now it was clear that there were many specific bacterial viruses; from the inception of the biochemical and biophysical study of their

---

[1] The viruses so far isolated from the blue-green alga *Plectonema boryanum* show some similarity to the bacterial viruses: in the electron microscope they appear as icosahedrons with short tails. Chemically, they appear to be DNA viruses (340).

86

behaviour, which developed into molecular biology and genetics, they have been known, collectively, as 'phage'.

The first paper describing the phenomenon was published in the *Lancet* on 4 December 1915 and, like other papers printed in that year, it bore the marks of difficult times and circumstances. Its author was F. W. Twort, and the title (451) gave no indication that it contained anything more than many other articles of the period dealing in a general way with the mysterious nature of ultramicroscopic viruses. It had occurred to Twort, a dedicated microbiologist, that just as non-pathogenic bacteria exist alongside the pathogenic varieties, so there might be non-pathogenic strains of viruses in addition to the obviously pathogenic ones. It was Twort's optimistic hope that such non-pathogenic viruses might be more easily cultivated than their pathogenic counterparts; it turned out to be a vain hope. He then attempted to isolate an organism which would reproduce vaccinia in animals, again with discouraging results; but he made an unexpected observation. Inoculating agar tubes with glycerinated vaccinia lymph which in some cases was contaminated with certain micrococci, he found colonies of white and yellow cocci undergoing surprising changes. Incubated at body temperature, these cultures would develop transparent areas which on examination by staining and microscopy could be shown no longer to contain intact bacterial cells, only 'minute granules, staining reddish with Giemsa'. Twort also showed that the transparent material at high dilutions passed through fine-pored porcelain filters, and that just one drop of this dilute filtrate would render an agar tube 'unsuitable for the growth of micrococcus'.

Twort went on to discuss the implications of his discovery. Twenty years later Eugène Wollman pointed out that Twort had at this early stage anticipated all the theories which had since been advanced concerning the nature of the phenomenon (480). Twort suggested that the 'transparent dissolving material' might be: (1) a stage in the life cycle of the micrococcus which will not grow on ordinary media, or (2) an enzyme secreted by the micrococcus, leading to its own destruction, or (3) an ultramicroscopic virus which might be considered an acute infectious disease of micrococci. The cautious and objective way

87

in which Twort presented these various theories has caused some commentators to dismiss his paper as nothing more than a first recording of a lytic phenomenon. In fact, as Wollman, ever a shrewd observer, noticed, Twort's short, compact paper showed remarkable breadth of vision. In view of the circumstances, his unwillingness to commit himself to any one unprovable hypothesis should be commended rather than criticized.

In the introduction to his paper, and in the final paragraphs, Twort made it clear that restrictions on time, funds and movement imposed by the war hindered further progress. He spoke wistfully of having missed an opportunity to study the problem in connexion with 'the dysenteric conditions in the Dardanelles', and concluded with the pious hope that 'others more fortunately situated' might proceed along the lines he had indicated.

Shortly afterwards, Twort, as captain in the Royal Army Medical Corps, was posted to the Base Laboratory at Salonika. According to the untraceable author of an unsigned obituary in the *British Medical Journal* (452), he lectured to fellow medical officers there on his newly discovered lytic phenomenon while the base was immobilized because of malaria, and among his listeners may have been a French medical officer called Felix d'Herelle. At this time, d'Herelle may or may not already, independently, have made similar observations;[1] after his return to Paris, in 1916, he was fortunate to witness an outbreak of dysentery such as Twort had hoped to encounter. In 1917 he could publish a paper on what he called an invisible microbe antagonistic to the dysentery bacillus (105). Whether or not he was aware of Twort's work, he did not refer to it. He wasted

[1] The dates given for d'Herelle's stay in Tunisia and his return to Paris, in the obituary written by Lépine (266), do not really lend support to the theory that he was in Salonika at the time suggested. On the other hand, there are also discrepancies between dates given by Lepine and by d'Herelle himself for his stay in Mexico where, he claimed years later, he had first observed '*taches vierges*' in cultures of coccobacilli isolated from infected locusts. Certainly mystery still surrounds the reasons why d'Herelle never mentioned these early observations in his paper of 1917, nor in the historical introduction to *Le bacteriophage* in 1921. He first referred to the locust phenomenon in a chapter added to the English translation of the book published in 1922 (107). It is a very short and spare account compared to the elaborate, almost lyrical description he produced for *Science News* years later in 1949, the year of his death (109). The conflicting evidence for and against d'Herelle's claim has recently been examined by D. H. Duckworth who has rightly concluded that the question must remain unsolved (122).

no time speculating on various possibilities with regard to the nature of the phenomenon; he was immediately convinced that the agent responsible was a living, microbial predator of the bacillus, and coined a name for it. The name stuck, and d'Herelle spent the rest of his life defending his position as the author of bacteriophage.

Careful examination of the two initial papers by Twort and by d'Herelle invites the conclusion that whereas d'Herelle immediately took up an intractable position and refused to consider any alternative to his initially formulated view of bacteriophage as tenuous, independently living and multiplying predators on the bacterial cell, Twort saw the wider possibilities and acknowledged that they could not be rejected without proof of a kind very different from that obtainable in the 1920s. On the other hand, d'Herelle followed up his initial paper on bacteriophage with many subsequent ones, most importantly the classic work, first published in 1921, *The bacteriophage* (106; and note 1, p. 88). In this volume he lucidly discusses many aspects of the phenomenon of bacteriophagy; and presents methods for quantitative determination, notably the plaque method, so useful in work with bacteriophage, and which was much later successfully adapted to work with animal viruses (cf. Chapters 6 and 9) (Fig. 19).

Twort had already hinted at the possibility of the therapeutic use of the 'dissolving substance', whatever its nature. D'Herelle and many others saw this as a very real prospect which fed a growing interest in the phenomenon in many quarters. For years hopes ran high that specific bacteriophages might be put to therapeutic use in the battle against specific bacterial diseases. Such hopes were never justified;[1] meanwhile, attempts to interpret certain inexplicable facets of the phenomenon in terms of its nature continued.

An alternative theory, considered by Twort and expounded by Bordet and Ciuca, claimed that the lytic phenomenon was due to an enzyme produced by the bacterium itself. It was a reaction mechanism akin to that proposed for tobacco mosaic

[1] This thwarted hope found its way into the literature of the period as early as 1925, when Sinclair Lewis' Martin Arrowsmith disastrously failed to control a plague epidemic by means of bacteriophage, and instead resigned himself to the study of its essential nature.

Fig. 19. D'Herelle's original presentation of 'Taches vierges' which was developed into the plaque counting method. On the far left an intact bacterial culture; no. 2 tube from the left has just been inoculated with phage; in no. 3 three lysed spots have developed; in no. 4 they are becoming confluent; and in no. 5, last tube on the right, complete lysis has taken place. From d'Herelle (1921) (106).

virus by Woods (489) and F. W. T. Hunger (223), although as far as this plant disease was concerned the enzyme theory was disproved by Allard (5) at about the time Twort's original paper appeared. Bordet was a formidable opponent for d'Herelle. Having just received the first post-war Nobel Prize for his contributions to immunology, notably his interpretation of the mechanism of immune response, he was clearly influenced in his attitude to bacteriophage by his own results in other areas; but from his first paper with Ciuca in 1920 (52), Bordet had a hunch that the phenomenon was somehow connected with heredity, more especially the processes of mutation. In time his hunch developed into a firm conviction, and Bordet took the stand that the phenomenon he called 'transmissible autolysis' was caused by an over-production of an enzyme by the bacterial cell which was thus producing the agent of its own destruction. Bordet described this as 'exaggerated physiological reactions' which in the course of the processes of mutation acquired a 'frankly pathological character', resulting in the lysis of perhaps less desirable mutants, so as to 'discipline the evolution of the species, by repressing the tendency of certain strains to obtain an excessive preponderance, and thus permitting the appearance or the survival of other varieties' (50).

Bordet was supported by the views of André Gratia, a younger staff member of the Institut Pasteur in Brussels in the 1920s. He was the first to rescue Twort from obscurity and to point out, in 1921, that Twort had observed and recorded the lytic phenomenon in 1915, two years before d'Herelle (180); and he countered d'Herelle's assertion that the phenomena were not identical[1] with experimental proof to the contrary.

The concept advanced by Bordet and by Gratia of a 'transmissible autolysis' due to an enzyme with the ability to induce its own reproduction by the bacteria it was lysing, was more compatible than d'Herelle's virus theory with the discovery of increasing numbers of lysogenic bacterial strains in the early

---

[1] Early in the 1920s d'Herelle had postulated the existence of two separate phenomena. The one discovered by Twort, wrote d'Herelle, should be called *bacterioclasis*; it consisted in a fragmentation which left the residue described by Twort as 'minute grains staining reddish with Giemsa'. The other then was his own discovery of *bacteriophagis*, in which he claimed lysis was complete to the extent that there was no residue whatsoever (108).

1920s. The complex nature of the phenomenon of lysogeny, made even more difficult by the existence of *carrier* strains as well as plain *lysogenic* strains of bacteria,[1] was a barrier to real progress in the study of bacteriophage for many years. While d'Herelle took the simple view that bacteriophage is a virus which infects sensitive bacteria, multiplies and lyses the host cell, experiments made between 1920 and 1925 by Bordet and Ciuca (52), Gildemeister and Herzberg (179), Otto and Munter (339), Lisbonne and Carrère (273) and Bail (21) gradually brought recognition that the behaviour of certain strains of *Escherichia coli* required a different explanation. In 1925, Otto Bail in Prague and Jules Bordet in Brussels both published their independent proof of the existence of strains of *E. coli* whose individual members would each produce a clone with the ability to generate phage even in the absence of free bacteriophage (21; 51).

This may be called the discovery of lysogeny – discovery, but not understanding. Bordet came no closer to understanding lysogeny than any of the others who worked with the lysogenic strains of *E. coli* in the 1920s; but his preoccupation with heredity as the 'continuation of a purely individual physiology' which merely became 'pathologically exaggerated' in the case of 'transmissible autolysis' brought him close to a correct interpretation of lysogeny without in any way understanding it. He wrote: 'The theory of autolysis implies accordingly that the sensitive bacteria already possess materials necessary for the formation of this principle, before they have been brought into contact with it. . .' He thought the ability to produce phage was inherent in the heredity of the bacterium, or even 'inserted' in its normal physiology; but at the same time he stated in so many words that he did not consider bacteriophages as 'materialized hereditary properties'. Analogy with earlier discoveries of his own led Bordet further astray '. . .There exist, moreover, in physiology, examples which one could think of in this connection, notably that of the agent, thrombin, responsible

---

[1] A *carrier strain* is a mixture of bacteria and bacteriophage in which the majority of bacteria is resistant, but sensitive variants occur which upon infection will show replication of phage. In a *lysogenic strain*, on the other hand, each bacterium is lysogenic and perpetuates the ability to produce phage.

for the coagulation of blood which, in acting, brings about its own reproduction. One of the constituents of thrombin, serozyme, exists in the plasma in the form of an inactive mother-substance which I have called 'proserozyme"' (50). The analogy was false; but when André Lwoff finally solved the riddle of induction in lysogeny (cf. Chapter 9), he called the structure carrying the potential for phage production in lysogenic bacteria 'prophage'.

The debate between d'Herelle, Twort and Bordet continued throughout the 1920s and into the 1930s. It was basically the old argument concerning the border areas between living and lifeless matter.[1] The comparison with filterable viruses was never far from the surface of the bacteriophage argument; but the difficulty pinpointed by Twort right at the beginning in 1915 remained. In the words of Twort: 'In the first place, we do not know for certain the nature of an ultra-microscopic virus'. And the difficulties were compounded by the erratic behaviour of the lysogenic bacteria. But if lysogeny presented a major stumbling block for the early phage workers, it was also the characteristic which in the right hands had much to contribute to the study of viruses and genes as it developed into molecular biology and genetics. In the words of Lwoff: 'Lysogeny occupies a privileged position at the cross roads of normal and pathological heredity, of genes and viruses' (291). By the time Lwoff was able to make this statement in 1953, there had been a profound change in many basic concepts in bacteriology, virology, genetics and general biology.

In the years leading up to World War II, activities in certain other areas of the natural sciences began to move inexorably towards work with viruses and bacteriophages. Techniques were improving in biochemistry, and even some physicists were slowly beginning to turn their attention to biological phenomena. Their immediate concern was with that central problem of the great question of the origin and the nature of life, the constitution of the genetic determinants. If an attempt were to

[1] Discussing this very problem in 1928, Boycott wrote: 'A good many people are willing to believe that the bacteriophage is generated by its own bacillus – which is probably the truth. And they would explain the way in which each bacteriophage more or less fits its own bacillus by its having originated from that bacillus...' (54; cf. p. 153).

be made to apply the laws of the exact sciences to the building blocks of biological material, then evidently the borderline areas of viruses and bacteriophage would be a logical place to begin.

Intimations of this approach were there at an early stage. The first had appeared on the eve of World War I; neither their authors nor their readers could then fathom the implications. Mrowka (315) and Andriewsky (15) had suggested that the agent of fowl plague might be of the nature of a globulin (cf. Chapter 4). The partnership of Peyton Rous and J. B. Murphy disagreed on the identity of the agent of a transmissible fowl sarcoma; while Rous called it a virus, Murphy referred to it as a 'transmissible mutagen' (320). Sanfelice (394) identified fowl pox virus as a nucleoproteid (sic). When bacteriophage received wide publicity because of d'Herelle's papers in 1917 and again in 1921, the end of the war provided breathing space for considerations of this esoteric subject.

The great Drosophila geneticist H. J. Muller was the first to observe a possible analogy between bacteriophage and gene, and to perceive that such an analogy would open up a whole new field of vision in genetics (68). In a remarkable paper read in Toronto in December 1921 (316), Muller presented aspects of his concept of genes and variation, and in the concluding paragraphs of this paper he turned to the newly discovered 'd'Herelle bodies' and their possible importance for the study of genetics. He wrote:

If these d'Herelle bodies were really genes, fundamentally like our chromosome genes, they would give us an entirely new angle from which to attack the gene problem. They are filterable, to some extent isolable, can be handled in test tubes, and their properties, as shown by their effects on the bacteria, can then be studied after treatment. It would be rash to call them genes, and yet at present we must confess that there is no distinction known between genes and them.

The theme was taken up shortly afterwards by Duggar and Karrer Armstrong (123) who saw that a similar argument might apply to tobacco mosaic virus; so that, right from the beginning, emerging molecular biology focused on bacteriophage and tobacco mosaic virus. Duggar and Armstrong wrote of the possibility:

That the causal agency in mosaic disease may be, in any particular case, a sometime product of the host cell; not a simple product such as an enzyme, but a particle of chromatin or of some structure with a definite heredity, a gene perhaps, that has, so to speak, revolted from the shackles of co-ordination, and being endowed with a capacity to reproduce itself, continues to produce disturbance in its path and 'stimulation', but its path is only the living cell [cf. Chapter 10 of this volume].

In Europe there were also biologists who saw the discovery of phage in a wider context. Beijerinck, who had not commented in writing on tobacco mosaic virus since the turn of the century, celebrated the occasion of the centenary of Pasteur's birth in 1922 with a few pages on Pasteur's part in the development of 'ultra-microbiology' (38). He included an evaluation of d'Herelle's discovery, and commented on current calculations concerning orders of magnitude: 'The dialysis test shows in my opinion very clearly that *Bacteriophagus* is of the same order of magnitude as a molecule of protein, and that the term *"contagium vivum fluidum"* which I long ago used of the virus of tobacco mosaic disease, is a fitting one'. As for d'Herelle's assertion that *'l'infiniment petit est aussi concevable que l'infiniment grand, nous n'avons pas le droit de lui assigner une limite'*, Beijerinck begged to differ, suggesting that 'inconceivable' would be a better adjective, thus showing himself to be in agreement with the thoughts and calculations on this subject expressed by McKendrick and by Errera 20 years earlier (cf. Chapter 7).

Inside the Institut Pasteur in Paris there were those also who saw that the uncompromising positions taken up by d'Herelle, on the one hand, and Bordet and his followers, on the other, were not likely to provide the final answers to the problems surrounding bacteriophage. Eugène Wollman was born in White Russia, and after studies at Liège arrived in Metchnikoff's laboratories at the Institut Pasteur in Paris in 1910. From 1920 onwards, he and his wife, Elisabeth, made bacteriophage their life work. Wollman's first note (478) on the subject was directly derived from the first paper by Bordet and Ciuca, in 1920 (52). Wollman agreed with their interpretation of a variation in the bacterial cells which was both hereditary and contagious – these first papers were still influenced by Darwinian ideas on pangenesis, and Wollman in 1920 likened Bordet and Ciuca's 'in-

95

tracellular factors' to Darwin's 'gemmules'. But these were also the years when men like Morgan and H. J. Muller changed the face of genetics, and Wollman soon adapted his ideas to more up-to-date concepts. By 1925 he made his position clear. Instead of partisan support for either d'Herelle or Bordet, he proposed a hybrid theory of his own, according to which 'the active principles of bacteriophages are particularly stable elements of cellular origin and act as material basis for certain hereditary characters – hereditary characters in the Mendelian sense' (479; 480). The ensuing phenomenon Wollman called 'hereditary contagious autolysis'.

Like Twort, Wollman was uneasy about his terms of reference. In 1925 he wrote:

The phenomenon discovered by Twort and d'Herelle might well, under the circumstances, mark a radically new departure. It was natural, at the beginning, to attempt to fit it into the known framework and to include bacteriophages, as did d'Herelle, with the filterable viruses. The question now arises if we ought not instead to reverse our relative values and consider the latter and their manifestations in the light of the notions suggested by the study of bacteriophagy. (479)

Ten years later, his ideas were bolder and more precise:

The idea of a material basis for characters, which under the name of the theory of genes has proved so fertile in the study of normal and pathological heredity in the higher organisms, thus seems capable of extension to certain processes of an infectious nature which up to the present were usually attributed to genuine viruses. (480)

In Australia, Burnet and McKie, working with a permanently lysogenic strain of Salmonella concluded that each bacterium must carry potential phage, i.e. a precursor or 'anlage' (64). When they disrupted their lysogenic bacteria by means of unrelated extrinsic phage, they recovered no infectious particles of the kind produced by their original lysogenic strain. The sum of their experiments allowed them to conclude that lysogenic bacteria do not contain infectious particles, and that phage is produced only as the result of a special process of activation. In a 1934 review article (59) Burnet wrote: 'One is almost forced to postulate that each lysogenic bacterium carries in intimate symbiosis one or more phage particles which multiply by binary fission *pari passu* with the bacterium.' With different collabor-

96

Fig. 20. Eugène Wollman. (Reproduced by courtesy of Dr E. L. Wollman, sub-director of the Institut Pasteur, Paris.)

ators, Burnet in the 1930s established a number of important facts which formed a solid foundation for further work on bacteriophage; in addition to those mentioned above, his work on surface and adsorption phenomena, indicating an analogy between the virus–host cell reaction and the antigen–antibody reaction, was of particular value.

During the same period, improvements in biochemical techniques produced, for the first time, indications of the chemical composition of bacteriophage. Until 1933, the only facts established concerning physical and chemical properties of bacteriophage had been measurements of size, based on ultrafiltration and ultracentrifugation experiments, primarily in the laboratories of Elford and of Bechhold.[1] About 1930, Max Schlesinger joined Bechhold's group; he soon proved himself the possessor of an independent and highly original mind. By improved physico-chemical methods, Schlesinger obtained important information about the dimensions of phage particles. Most important of all, he developed a method of differential centrifugation and filtration which allowed him to purify phage in sufficient amounts for chemical analysis (401). By such means he was able to show, before his tragically premature death in exile in London (403), that his pure phage material consisted of approximately equal amounts of protein and DNA (402).

By 1937, Schlesinger was dead, and Burnet was turning his attention to animal viruses rather than phage. In Paris, the Wollmans were moving closer than ever to a final understanding of the phenomena involved in lysogeny. They could establish, through study of the kind of lysogenic and non-lysogenic strains of *Bacillus megaterium* which had defeated den Dooren de Jong[2] (an erstwhile assistant of Beijerinck's), that new infective phage particles develop from non-infective precursors and are not direct descendants of pre-existing phages, i.e. bacteriophage is not actually present in a lysogenic bacterium, but is maintained in a potential, hereditary form (481; 482). On the eve of World War II, the Wollmans saw that in order to make further progress

[1] Elford reviewed both his own, Bechhold's and related work on several occasions, see, for example, 'Ultrafiltration', 1928 (137), and 'The size of viruses and bacteriophages, and methods for their determination', 1938 (139). See also Chapter 2.

[2] L. E. den Dooren de Jong discovered lysogeny in *Bacillus megaterium* in 1931 (119). Armed with the knowledge that phage was killed at certain temperatures, den Dooren de Jong concluded that no phage could have been present in the spores, and that consequently the new phage particles must have been produced by the bacteria.

Fig. 21. Memorial plaque to the Wollmans, unveiled at the Institut Pasteur, Paris in 1959. (Reproduced by courtesy of Dr E. L. Wollman, sub-director of the Institut Pasteur, Paris.)

they would have to be able to study the behaviour of single bacteria in isolation. They obtained a micro-manipulator, an instrument which became an essential part of the equipment with which Lwoff and also Dulbecco later achieved epoch-making results.[1] For many European scientists, World War II meant severe, but temporary, curtailment of important and promising experimental work. For the Wollmans, the interruption was final and irreversible. In December 1943, they were arrested in occupied Paris and soon afterwards deported to Auschwitz. They were never heard of again (483) (Fig. 21).

There were no immediate successors ready to take over where Burnet, Schlesinger and the Wollmans had left off. The phage work which evolved in so striking a way during and immediately after the war had different origins.

[1] Lwoff recorded the Wollmans' acquisition of a micromanipulator in 'Lysogeny' in 1953 (291), and his own use of the instrument in 'The prophage and I' in 1966 (293). Dulbecco referred to the manipulator in 'From lysogeny to animal viruses' in 1971 (129).

99

9. The bacterial viruses II

Our understanding of biological mechanisms at the cellular and molecular level has improved dramatically within a surprisingly short span of time after the close of World War II. Much of the information obtained has resulted from the work of what has come to be known as the 'phage group', a heterogeneous body of physicists, medical biologists and geneticists studying bacterial viruses.

Rarely, if ever, has the development of a scientific discipline been so well documented and chronicled even as it evolved. Numerous essays and several books dealing with aspects of the work, with the discoveries and the personalities of those who made them have appeared[1] since molecular biology and genetics first began to yield their impressive results.

This second phase of bacteriophage studies, and at the same time the transition from classical genetics to its molecular counterpart, had its origins in pre-war Berlin in the 1930s. In the then Kaiser Wilhelm Institutes – the nucleus of the admirable multitude of institutes which today flourish all over West Germany under the name of Max Planck Institutes – began at this time what is now taken for granted as a means of obtaining faster and more wide-reaching results in many fields, interdisciplinary teamwork. Initially, it centred around the man whose contribution never reached the editors of *Phage and the origins of molecular biology* (67), the Russian geneticist N. W.

---

[1] In addition to the two major collections of essays, *Phage and the origins of molecular biology* (67) and *Of microbes and life* (see 129), there are the two excellent works of reference, Gunther Stent's *Molecular biology of bacterial viruses* and R. Olby's *The path to the double helix* (335). Among countless shorter and more specialized contributions may be mentioned Lwoff's 'Lysogeny' (291) and 'The concept of virus' (292), and in a more or less sarcastic vein, Burnet's 'Men or molecules' (63) and Chargaff's 'Preface to a grammar of biology. A hundred years of nucleic acid research' (77).

Timoféeff-Ressovsky.[1] Then in his early thirties and working in the genetics division of the institute devoted to brain research, this man of vision was the natural leader of a group of young scientists who were interested in applying the laws of physics and mathematics to biological problems, and in particular those concerning the hereditary material. With K. G. Zimmer, 10 years his junior and working in the radiation department of the Cecilienhaus hospital in Berlin, Timoféeff-Ressovsky was attempting to learn something about primary physico-chemical processes taking place during genetic changes induced in *Drosophila* by ionizing radiation. In 1932, they were joined by Max Delbrück, who had come to Berlin ostensibly to work on radio-activity with Lise Meitner in Otto Hahn's institute.

His express hope was that the geographical closeness of the various Kaiser Wilhelm Institutes would provide him with an opportunity for contact with biological problems. He was not disappointed. Within a very short time, he was absorbed into the circle around Timoféeff-Ressovsky, and actively collaborating with him and with Zimmer. What began as informal discussions culminated in 1935 with the publication of the famous *Dreimännerwerk*, or 'green paper',[2] 'Über die Natur der Genmutation und der Genstruktur' (447).

It is a paper distinguished not only by its contents, but also by the arrangement of the contents. Unlike later examples of similar teamwork, its three authors did not collaborate to the extent of integrating the expression of their thoughts on paper – each wrote his own discrete part of the article. Timoféeff-Ressovsky dealt with the genetics, Zimmer with the target theory as developed by Dessauer (104) and Crowther (96), while Delbrück launched a theoretical physicist's model of gene mutation.[3] Only in the concluding pages of the paper did the three authors join forces to explain their theory resulting from their collaboration concerning gene mutation and gene struc-

[1] Timoféeff-Ressovsky returned to the USSR at the conclusion of World War II, and communication with Western colleagues appears to have been somewhat difficult in later years.
[2] This name quite simply refers to the green colour of the cover of the issue concerned of *Nachrichten der Gesellschaft der Wissenschaften in Göttingen*.
[3] Delbrück concluded that the view of gene mutation as an elementary process in a quantum mechanical sense, in particular as a definite change in complex atomic bonding, was qualitatively justified in the light of data from mutation experiments. The individual contributions of the *Dreimännerwerk* are discussed in detail by Olby in *The path to the double helix* (335).

ture. For Max Delbrück, this paper marked a permanent change in direction. An interest in biology, first kindled in Niels Bohr's institute in Copenhagen, took on an ever more attractive hue.

When the 'green paper' was published in 1935, time was running out for the Kaiser Wilhelm Institutes. Many of the most valued members of their staff left the country, or were forced to leave, for political reasons.[1] The war further disrupted the work of the Institutes. But thousands of miles away, in another continent, largely unaffected by the European war save for the welcome addition of highly gifted scientific refugees to its staff, was an institution which in its own way fostered the principles of interdisciplinary collaboration. But unlike the Kaiser Wilhelm Institutes which were scattered over a large area of Berlin and its suburbs, the California Institute of Technology at Pasadena was a small integrated institution, its disparate departments nestling together in comfortable isolation in the foothills behind Los Angeles. Within its walls it was deliberate policy to encourage interdisciplinary exchange of ideas as much as within the different Kaiser Wilhelm Institutes. Here the genetics of T. H. Morgan's biology division, the protein chemistry of Linus Pauling's chemistry division, and the combined efforts of physicists and physicians to elucidate the effects of X-rays on human cancers, co-existed to mutual advantage.

In 1937, Delbrück left Berlin for Pasadena, to pursue his interest in *Drosophila* genetics in T. H. Morgan's department. He took with him a brief memorandum, written shortly before, which he had called 'Riddle of life' (see 101). Influenced by Stanley's claim to have crystallized a protein molecule with the properties of tobacco mosaic virus (424), it summed up the prevailing attitude of men such as Niels Bohr, Timoféeff-Ressovsky, H. J. Muller and Delbrück himself, to the problem of its title. It concluded that a study of virus replication, supposing viruses to be molecules, would appear to be a promising means of obtaining clues to gene replication. On arrival in Pasadena, Delbrück discovered that a filterable virus was indeed assuming general importance as a model system; but it was not tobacco mosaic virus.

[1] The effects of this forced move are discussed in *The intellectual migration: Europe and America 1930–1960* (157). In an essay called 'Emigré physicists and the biological revolution' Donald Fleming of Harvard also perceptively examines the *Dreimännerwerk* and its ramifications (156).

The bacterial viruses II

The choice of model system at Pasadena was the outcome of biochemical studies on malignancies; because of the transmissibility of some animal tumours by cell-free filtrates, a filterable virus seemed a logical choice. Requiring less outlay in terms of time, equipment and space – no experimental animals or even plants, only plates of bacteria – bacteriophage became the chosen model system at the California Institute of Technology. Above all, the biologists there made their choice of phage on the basis of a comparison of three processes, i.e. the fertilization of an egg cell, the infection of a bacterium by a phage particle and the virus-induced conversion of a normal cell into a malignant one (141). The obvious analogies and differences between the three phenomena give rise to the hope that an understanding of the mechanism of bacteriophagy would provide insight into the virus-induced malignancies.

Delbrück immediately joined in the bacteriophage studies, at first with E. L. Ellis. They began by studying growth curves for coliphages, designing the one-step growth experiment[1] which led them to the individual burst, and to the realization that the average number of phage particles released varied from a few to well over a hundred in a single burst (142). For Max Delbrück, as for the majority of his later associates, phage was not primarily a means of exploring virus behaviour and replication; it was, above all, the chosen tool for an examination of genetic processes, and a possible clue to the nature of the gene itself. That had been his aim since his early days with Timoféeff-Ressovsky, and perhaps even before that, since he listened to Niels Bohr's lecture on 'Light and Life' (48) in Copenhagen in 1932, before he went to Berlin (see 335). Gunther Stent has suggested (426) another motive at the back of the physicist's mind: the 'romantic' search for a paradox which might involve the discovery of 'other laws of physics'.[2] Whatever his motives – and probably the motives of any scientist worth his salt are likely to be far more complex than he himself is apt to realize when he sets out to solve an individual problem – the impact of his work on phage profoundly affected the post-war world of biology.

[1] This type of experiment is based on the simultaneous infection of a large number of cells. For the purpose of measurements and interpretation it is assumed that all cells are infected, and that identical processes take place in them all at the same time so that the whole system may be treated as one large cell.
[2] This argument has been taken a step further by Lewontin in an essay review of *Phage and the origins of molecular biology* in 1968 (270).

After two years in Pasadena, Delbrück left to teach physics at Vanderbilt University; but as far as research was concerned, he was totally committed to work with phage, and actively looking for converts to his cause. By now, Europe had gone to war, and in the United States there was a renewed influx of young scientists from countries occupied or otherwise threatened by the German war machine. In 1940, Salvador Luria came to work in the department of surgical bacteriology at Columbia University in New York. A graduate of the medical school in his native Turin, he had spent two years in Paris, where he had learned the techniques of bacteriophage studies from Eugène Wollman himself; at Columbia he worked with Exner on effects of X-rays on bacteriophage (288). Not long after his arrival in the States, he met Delbrück at a meeting in Philadelphia. They immediately realized their common interest in phage, and acted on the discovery with what Luria later described as a '48-hour bout of experimentation' (284) which might perhaps be seen as the first tiny nucleus of the phage group. It was really a session for Luria to instruct Delbrück in the plaque technique he had learned from Wollman in Paris – a legacy from the old school of bacteriophage work to the new. But unlike the early phage workers, who pursued their studies in almost total and deliberate isolation from each other, the phage group grew from an initial affinity of mind and purpose between Delbrück and Luria into a team on an unprecedented scale, which from its inception had great unity of aim and made a point of full consultation all along the line.

From this their first meeting, it was obvious that Delbrück and Luria had a common goal, and a similar philosophy for its attainment. They also made a conscious effort to attract as many others as possible to their cause. The phage group did not grow in a haphazard way – it developed steadily through the deliberate recruitment by its founder members of eligible geneticists and physicists with common interests. From 1945 onwards, Delbrück organized an annual 'phage course' at Cold Spring Harbor.[1] But no special efforts were made to attract chemists

---

1 In a preface to *Phage and the origins of molecular biology* the editors called Cold Spring Harbor 'a small station of meager [sic] resources on the shores of Long Island Sound' and acknowledged the debt of the phage workers to the 'imagination and enterprise' of its director from 1941 to 1960, Milislav Demerec (67).

or biochemists. It was Delbrück's express contention that a chemical approach could contribute little of value to their study of genetic determination; and in spite of his medical background, Luria was no more enthusiastic than Delbrück. However, a few biochemists eventually became accepted members of the group. One was S. S. Cohen, whose training in biochemistry had provided a suitable background for work with Stanley on the molecule of tobacco mosaic virus (87). Turning to bacterial viruses, he attended the second phage course organized by Delbrück at Cold Spring Harbor in 1946. Very soon, he could show the merits of the application of modern quantitative biochemistry to the study of the phage–bacterial host system (85; 86).

The recruitment of physicists to work in biology and in particular genetics was also boosted during the years of the war by the appearance of a small volume, written by Erwin Schrödinger in exile in Dublin.[1] Under the title *What is life? The physical aspect of the living cell* (405), it was a physicist's attempt to come to terms with the processes of living cells and heredity mechanisms via the established laws of physics. Delbrück's model was discussed at length. But the biological models chosen to illustrate his thesis came no lower in the scale of organization than *Drosophila* and the plant cells of *Tradescantia* and *Fritillaria*. At no stage did Schrödinger invoke viruses, much less comparisons between viruses and genes. There can be no doubt of the considerable impact of the book on the physicists who read it; among biologists, on the other hand, the effect appears to have been less pronounced.[2]

In Germany, on the eve of the war, it had for the first time become possible to obtain visual impressions of viruses, in the

---

[1] The most important period in Erwin Schrödinger's working life was the Zürich years (1921–27), when he laid the foundations of wave mechanics and made important contributions to quantum mechanics. After 1927, in Berlin, he (and Einstein and Planck with him) found it difficult to adjust their traditionally deterministic thinking to the new interpretations offered in Göttingen, Copenhagen and Cambridge. 'I can't imagine that an electron hops about like a flea' Schrödinger is reported to have commented (205). According to the same source, the ensuing disappointment may have influenced Schrödinger's later outlook.

[2] Although J. D. Watson, in 'Growing up in the phage group', describes himself as having become 'polarized toward finding out the secret of the gene' by reading *What is life?* after undergraduate training in more conventional natural history (470).

recently completed electron microscope.[1] Kausche, Ruska, and Pfankuch had first published electron micrographs of tobacco mosaic virus (240); in 1940 they added studies of bacteriophage (357; 388). In spite of the difficulties in communication at the time, the United States National Research Council lost no time in following up this lead, and with the financial and manufacturing backing of the Radio Corporation of America arranged for an investigation of the biological application of such an instrument. T. F. Anderson received a fellowship to explore the possibilities in 1940; a year later, he was approached by Luria, who was interested in checking estimates of phage sizes, which he had recently made with Exner, against phage particles as visualized in the electron microscope. Within a short time they had confirmed the German observations of 'sperm-shaped' particles; they also found the structures of the 'head' and 'tails' parts, respectively, to be characteristic of individual strains of phage. There was general agreement between the measurements they were able to make and those suggested by indirect methods (285).

Shortly afterwards, in the summer of 1942, Luria brought Delbrück along to the 'biologists' summer camp' at the Marine Biological Laboratory at Woods Hole, where an electron microscope had been installed for Anderson. Together the three of them studied the pictures and considered possible interpretations. One important question concerned the entry of virus into the host cell, prior to replication. There were to be no easy answers. Difficulties in preparation of samples, and in particular, distortion of pictures due to drying, supported the notion of sperm-like particles moving towards the host cell, head first and possibly propelled by the 'tail'. But what perhaps surprised the observers most of all was the total absence of phage particles inside the cell (see 287). All appeared to be adsorbed on the outside, but none, not even parts of one, were ever seen inside the cell. Did this mean that only one phage particle penetrated, in the process changing the permeability of the cell wall to the exclusion of any others? It would account for the phenomenon

[1] The first electron microscopes date from the early 1930s, but their early history is more than usually rife with claims and counterclaims to priority, see Marton's *Early history of the electron microscope* (309).

of interference between different virus particles attacking a cell, observed by Luria and Delbrück at about this time (102; 286); it would also support the sperm–egg analogy which had been a *raison d'être* for bacteriophage work in the first place, and which had not unnaturally been boosted by the apparent tadpole shape of the phage particles observed in the electron micrographs. Of course, as Anderson himself has pointed out, '...the analogy breaks down completely when we consider the products of the two reactions for phage-infected bacteria are lysed with the production of more phage while the fertilized egg has never been observed to break down with the production of more sperm' (9).

Such reflections did not deter Luria and Delbrück. They were intent also on laying the ghost of a certain precursor theory championed by A. P. Krueger. Krueger had been one of J. H. Northrop's assistants at the Rockefeller Institute in the early 1930s. Like Stanley, and Northrop himself (333), he had been much influenced by the successful work on enzymes there. If Stanley's initial identification of the crystals of tobacco mosaic virus as protein molecules appears in retrospect hasty (p. 123), it seems reasonable to assume that he was led astray by a preconception that they should conform to the pattern of the crystallized enzyme. In much the same way Krueger tried to formulate a theory according to which phage developed from a precursor, already present in the bacterium, in a reaction analogous to the conversion of pepsinogen and trypsinogen into the respective enzymes *in vitro* (252).

The one-step growth experiment had produced information on the overall nature of phage replication. The next question asking to be solved concerned the mechanism of replication inside the infected cell. It could be answered only by breaking open the host cell at intervals between the point of infection and the release of phage progeny by lysis. Doermann first successfully developed such a technique in 1948 (113; 114), using cyanide and large amounts of a different phage (T6 phage to lyse T4 infected cells 'from without'). The surprising result of these first experiments by Doermann was to show that no infective phage particles could be recovered from the infected cell for a period of 10 minutes following infection. After this initial period of apparent inactivity, described as an eclipse (283), the number of infective phage particles gradually increased until their release by spontaneous lysis of the host cell.

The difficulties in preparing specimens, which had left Anderson with ambiguous pictures offering no possibility of determining whether phage particles attached themselves to the host cell surface by their heads or their tails, were not resolved until 1949. It then became obvious that the phage particles were adsorbed on the cell surface by their tails, and that their basic structure was a proteinaceous coat containing DNA (Anderson's 'osmotic shock' experiments (8)). Furthermore, while still no trace of phage particles had ever been found inside the invaded cell, empty-headed phage 'ghosts' – the protein coat – could be observed on the surface (210). Anderson later recalled: 'I remember in the summer of 1950 or 1951 hanging over the slide projector with Hershey, and possibly Herriott, in Blackford Hall at the Cold Spring Harbor laboratory, discussing the wildly comical possibility that only the viral DNA finds its way into the host cell, acting there like a transforming principle in altering the synthetic processes of the cell' (9).

Within little more than a year, the 'comical possibility' proved to be the hard fact. Using another post-war tool of enormous impact across the whole spectrum of biological research, the radioactive tracer, Hershey and Chase (212) were able to show that the protein coat of the particles of the so-called coliphage T2 is adsorbed onto the surface of the host cell by its tail, which then acts as a syringe, injecting the DNA contents of the phage particle into the cell. It was elegant proof of a very important step in phage reproduction, although the authors remained cautious in their conclusion, writing: '...the protein probably has no function in the growth of intracellular phage. The DNA has some function. Further chemical inferences should not be drawn from the experiments presented.'

There was no doubt of the implication. The genetic material in phage reproduction was nucleic acid, not protein. It was a discovery which was in perfect accord with the earlier (1944) observation by Avery, MacLeod and McCarty (20) that DNA was the carrier of gene information in *Pneumococcus*, although

Hershey and Chase did not mention the fact.[1] It was also a discovery which did nothing to slow down the efforts of Watson and Crick to present a structure for DNA. Their results were published in *Nature* in April 1953 (471).

Nor were these the only important papers published by members of the phage group between 1950 and 1953. Suddenly, several areas of research within the field originally mapped out by Delbrück and Luria came to fruition all at once, including some which had developed tangentially. Renato Dulbecco had come from Turin to Indiana University while Luria was there. Absorbed into the phage group, he had gone out to Pasadena in 1949. During his time with Luria, he had reported the phenomenon of photoreactivation (125), which to the phage group was a disconcerting sequel to the discovery of ultraviolet inactivation of bacteriophage, darkening rather than illuminating their picture of ultraviolet mutagenesis. At the California Institute of Technology, Dulbecco's work was to change direction. The institution had received an endowment intended for the development of work on animal viruses in addition to the existing research on bacterial viruses. The challenge was eagerly accepted by Dulbecco. The necessary first step was the development of a quantitative approach comparable to the plaque-counting method enjoyed for so long by those who worked with bacteriophage. After some initial difficulties, and helped by concurrent advances in tissue culture techniques,[2] Dulbecco, in 1952, could present details of a plaque-counting method for use with animal viruses (126).

For all the valuable information gathered and the pooling of high-powered minds, there was one aspect of the problem of bacteriophage which was ignored by the phage group for years. D'Herelle had refused to consider the possibility that bacteriophage might exist in intracellular symbiosis with its bacterial host cell. Delbrück and his group had no reason to consider it; they happened to work with strains of T phages which did not

---

[1] The fate of the paper by Avery *et al.* and the reasons for the delay in general recognition of its importance have been discussed by H. V. Wyatt (490).

[2] Dulbecco later described the gradual development of his ideas in amusing detail (cf. note 1, p. 77). The use of trypsin digestion to prepare cell suspensions occurred to him during a visit to Enders who at the time was trypsinizing tissue fragments in poliovirus work (128).

exist in the lysogenic state. The final unravelling of the apparent paradox of lysogeny was to come, with poetic justice, from the institute where the Wollmans had laboured so long and so painstakingly, and had come so close to an understanding of the problem. André Lwoff, who, with the collaboration of younger colleagues, made the problem his own, had learnt about lysogeny from Eugène Wollman, although by his own account he was not sufficiently 'impressed' initially by the phenomenon (293). From the Wollmans to Lwoff, the Institut Pasteur in Paris has remained the centre of attack for the phenomenon of lysogeny. But however much the study of lysogeny may appear to have become a French prerogative, indirectly the influence of the phage group was at work here, too.

The second summer of the phage course at Cold Spring Harbor coincided with the resumption of the Cold Spring Harbor Symposia after the war, for the first time since 1942. The subject under review was 'Heredity and variation in microorganisms', and the participants included some who had weathered the storm in Europe[1] and who were now granted their first glimpse of the promised land of American post-war scientific endeavour. From the Institut Pasteur in Paris came Jacques Monod, Raymond Latarjet and André Lwoff. The latter was there because of an interest in the genetics of *Moraxella*. He did not attend the phage course; but the impact of introduction to what he (and Monod) later called 'the powerful phage church' was considerable, both materially and intellectually.[2] In any case, during the next couple of years, back in Paris, both Lwoff and Monod were somehow drawn to work with bacteriophage. By 1949 they were entirely caught up in this interest.

Unlike the American phage group, Lwoff was careful to select

[1] Of the three 1965 French Nobel Laureates, André Lwoff received the Medaille de la Resistance in 1946, while Jacques Monod, eight years his junior, commanded an underground unit of the French Resistance after the fall of France and earned the Croix de Guerre. The youngest of the three, François Jacob, had not completed his medical studies at the outbreak of war. Having escaped on one of the last boats to leave for England in 1940, he joined de Gaulle's Free French forces, fought throughout the African campaign, and was seriously wounded. He holds the Croix de la Liberation.

[2] Lwoff's entertaining account of the 'hungry Frenchmen's' encounter with the stimulating academic climate of Cold Spring Harbor plus the sometimes surprising gastronomic novelties ('red beans seasoned with peanut butter') is preserved for posterity in 'The prophage and I' (293).

a lysogenic strain of *Bacillus megaterium*.[1] In a series of elegant experiments in 1949 and 1950 he was able to show, with his colleagues, that lysogenic bacteria do not contain virus particles, and that the inherited capacity to produce bacteriophage is propagated only as a potential character (294). In other words, what is reproduced by a lysogenic bacterium is the genetic material of the phage which Lwoff called prophage.[2] Prophage then is a non-infective, vegetative entity which upon induction by external factors enables (or compels) the bacterial cells to produce infective phage particles, instead of its own products, resulting in the ultimate lysis of the cell. The first factor shown to cause induction was ultraviolet light (295); later experiments proved that lysogenic bacteria treated with hydrogen peroxide, X-rays and certain chemicals were similarly induced.

And so at last a picture had emerged of the general mechanism of lysogeny (Fig. 22); but the genetics of the phenomenon invited further exploration. At the Institut Pasteur, the task fell to the younger Wollman.[3] His first results (484) as well as results obtained simultaneously and independently by E. and J. Lederberg (263) were somewhat unclear; different crosses between lysogenic and non-lysogenic bacteria gave different results. In some cases there appeared to be some linkage between the lysogenic character (as determined by the λ prophage of *Escherichia coli*) and other characters determined by the bacterial genes; in other cases there was none.

At this stage, analysis of the genetics of lysogeny became of necessity inextricably linked to analysis of the genetics of the bacterial cell, and positive results had a bearing on both. In the early 1950s Wollman began a long and fruitful partnership with François Jacob.[4] Their work, and that of W. Hayes in London

---

[1] What to Eugène Wollman and den Dooren de Jong was *Bacillus megatherium* has now, after later taxonomic revisions, become *B. megaterium*.

[2] Cf. Burnet and McKie's 'anlage', Chapter 8 (64).

[3] With some amusement the Wollmans' son, Elie L. Wollman, now a sub-director of the Institut Pasteur, has described how, in a Caltech library file in the late 1940s, he came across a reference to his parents' papers of 1937–38 in which they concluded that phage entered a non-infectious phase upon entry into the bacterial cell. The file card bore the laconic comment: 'Nonsense' (485).

[4] His medical studies interrupted by his war experiences (note 1, p. 110), François Jacob after demobilization tried out a number of jobs, including one in the pharmaceutical industry. When he finally decided to return to the biological sciences in Lwoff's department, he found entry difficult, and perhaps only ultimately made possible when Lwoff's discovery of the induction phenomenon came to his aid (230).

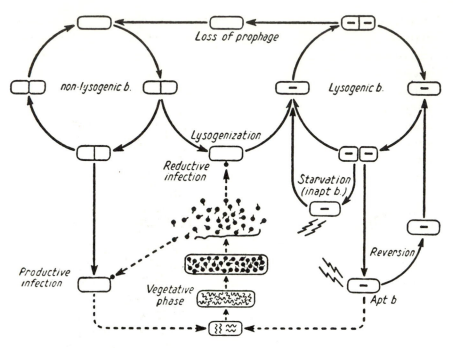

— *Prophage*, ∿ *Gonophage*, ● *Bacteriophage*, ∿ *Inducing agents, b bacteria*

Fig. 22. Lwoff's (1953) diagram explaining the processes of lyso-geny. From *Bacterial Review* (291). (Reproduced by courtesy of Dr A. Lwoff.)

and of Lederberg and his colleagues, yielded throughout the 1950s an impressive amount of information regarding the gene-tics of bacteria and of bacteriophage. Conjugation and genetic recombination had been discovered in the K12 strain of *E. coli* in the mid-1940s by J. Lederberg and E. L. Tatum (264); in 1951, Esther Lederberg showed that this strain is in fact lysogenic, carrying a phage she called lambda (262). With this system it finally became possible to attempt an analysis of the genetics of lysogeny, while almost simultaneously the discovery of the phenomenon of transduction by Zinder and J. Lederberg in 1952 (494) suggested another mechanism, in addition to con-jugation and transformation, by which the genetic factors of two different bacterial cells could be combined in one offspring cell, in this case by the transfer of DNA from one cell to the other through the intervention of a phage particle.

At the same time, the Lederbergs in America and Elie Woll-man in Paris were putting the λ phage to work. The first aim was to determine the distribution of lysogenicity among crosses between lysogenic (λ phage) and non-lysogenic strains of *E. coli*. The results were not conclusive; a better understanding of the phenomenon of conjugation was needed. To gain such an understanding, Elie Wollman proceeded to use a blender, the kitchen tool which had achieved fame in the experiment by Hershey and Chase (212). In this way Wollman and Jacob were able to interrupt conjugation at selected points, and thus to show that the male of the bacterial pair slowly injects its chromosomes into the female (486). When the cells were separated, by the blender, during conjugation, at a point where the male chromo-some was broken during penetration, they found that the fragment which had already entered the female cell was able to express its genes and undergo recombination on its own.

With such improvements in methodology it became possible to gain a better understanding of bacterial and phage genetics. The bacterial chromosome was found to consist of a single, formally circular linkage group to which it was possible to add or subtract so-called episomes, i.e. other genetic elements such as prophage. According to Jacob and Monod (231), prophage in the lysogenic cell is controlled by a *repressor* present in the cytoplasm and is itself formed under the influence of the pro-phage, that is, as the expression of one gene of the prophage. The phenomenon of *induction* is then supposedly caused by a transient reduction in the level of the concentration of the repressor, allowing initiation of the phase of vegetative phage replication. Consequently, ultraviolet light and other factors cause induction by destroying or inhibiting the repressor. The difference betwen *temperate* and *virulent* phage then appears to be the ability of the temperate phage to produce its own repressor.

The experiments and hypotheses of Monod and Jacob at this time also opened up a central question concerning protein synthesis in the cell. It had been confirmed in 1956 (493) that proteins are made on the ribosomes in the cytoplasm; but how the genetic information was carried from nucleus to cytoplasm was unknown, and the available facts did not altogether support the 'one gene–one ribosome–one protein' theory.[1] Instead, Jacob and Monod proposed a less paradoxical alternative; i.e. that ribosomes are non-specialized structures, and that information is carried from gene to ribosomes not by ribosomal RNA, but by an unstable intermediate, a 'messenger RNA'. The validity of this hypothesis was proved by Jacob, in collaboration with Sydney Brenner from Cambridge and Meselson in Pasadena, in an ingenious experiment based on the differential use of 'light' and 'heavy' isotopes in combination with density gradient centrifugation of purified ribosomes. By such means they were able to show that no new ribosomes were formed following phage infection, and that most, if not all, protein synthesis in the infected cell took place in pre-existing ribosomes. They also found that upon infection with phage the cell synthesized a new, relatively short-lived, RNA with a base composition corresponding to that of the phage DNA, and that this new RNA became attached to the existing ribosomes (55). As for the genetic material of the bacterial virus itself, Hershey and his colleagues have shown in the 1960s that (1) virus particles contain single molecules of nucleic acid; (2) the molecules are species specific; and (3) different viral species contain nucleic acids which differ in length and nucleotide sequence (211). Through the work of Hershey it finally became possible to pinpoint the essential difference between viruses and bacteria and to arrive at a definition untrammelled by the negative criteria used since the turn of the century. A virus then may be defined as an organized particle containing *either* DNA *or* RNA but not both, and reproducing itself only from this, its own indigenous genetic material. Its molecule of nucleic acid can replicate, express its genes, form virions, only within a living cell, by using the existing mechanisms (enzymes, etc.) of that cell. Beijerinck had no way of knowing of DNA and RNA; it is worth recalling here the words he first wrote in 1898, of tobacco mosaic virus: '. . .might conceivably serve as an explanation that the contagium, in order to reproduce, must be incorporated into the living protoplasm of the cell, into whose reproduction it is, in a manner of speaking, passively drawn' (36).

[1] In the light of their newly gained insight Jacob and Monod in 1961 concluded that 'Structural genes obey the one gene–one protein principle, while regulator genes may affect the synthesis of several different proteins' (231).

Nor are the prophetic words by Beijerinck the only link between the high sophistication of present-day molecular genetics as practised by the phage group workers (and the growing body of their students and followers) and the origins of the virus concept in the late nineteenth century. Ivanovski was clearly influenced in his attitude to the filterability of the pathogen of tobacco mosaic disease by the discovery of diphtheria toxin (227). Long after vaccination had removed diphtheria from the list of major concerns in pathology, the phage group found their molecular genetics developing links with this bacterial disease.

Corynebacterium diphtheriae was first described by Klebs in 1883 (242); a year later Loeffler demonstrated its role in the pathogenesis of the experimental disease (277). When, in 1888, E. Roux and A. Yersin discovered its toxin (386), and Behring in 1890, the anti-toxin (35), it became possible to control the disease. But already Loeffler had reported that he could sometimes isolate from the throats of normal, healthy individuals organisms which were morphologically indistinguishable from the pathogen; in animals they were avirulent. From time to time in subsequent years there would be reports of the appearance of 'avirulent' diphtheria bacilli in healthy children and animals, but little was known about the difference between virulent and avirulent strains until 1951. Diphtheria had by then ceased to present much of a public health problem; its aetiology now achieved renewed, if academic, interest. It was discovered by V. J. Freeman that the toxigenicity of the virulent strains of *C. diphtheriae* depends on the carriage of specific bacteriophage (162). This intriguing new aspect of an old problem was subsequently studied by Groman (182; 183) and by Barksdale and Pappenheimer (25) in the light of other discoveries made during the same decade in the areas of lysogenicity, lysogeny and phage–bacterial chromosome relationships. The mechanism of toxin production and its interaction with the metabolism of the host bacterium could clearly have implications outside the narrow field of *Corynebacterium*. The pathologists who lost interest in the lytic phenomenon when it proved of no therapeutic value may find themselves compelled to take another look at bacteriophage.

10. Plant and insect viruses

If the previous chapter has left the reader with the impression that the bacterial viruses far outshone all other viruses as models in structural and functional studies in general virology as it developed during the 1940s, it must be stated right away that the plant virus which gave rise to the concept has run them a very close second and provided a necessary complement. The prophetic words of Beijerinck (cf. Chapter 3) may have gone largely unheeded for many years, but tobacco mosaic virus continued to have a distinguished history of its own which reflected, and often preceded, the general development of ideas concerning mechanisms of viral infections.

If Beijerinck pre-empted the modern concept of the virus in his first paper on tobacco mosaic disease, another main contender among the hypotheses concerned with the nature of the filterable viruses was not long in following. It was published in the same journal in which the German version of Beijerinck's article had appeared, and its author was an American plant pathologist, A. F. Woods (489). Woods believed tobacco mosaic disease to be a pathological condition due to an accumulation of oxidizing enzymes induced by unfavourable external conditions. It was a theory which had elements of humoral pathology and even could bring to mind the perennial claims of spontaneous occurrence of rabies in dogs subjected to stress of various kinds – e.g. starvation, excessive heat, extreme thirst, etc.[1] Woods made no attempt to account for the apparently limitless transmissibility of the disease. His theory and a similar one expounded by Heintzel in Erlangen in a thesis submitted in the

[1] For rabies such claims continued to be made into the 1930s (487) having made regular appearances in the literature throughout the centuries (see 3; 445).

following year (203) were criticized by Hunger in 1905 (222; 223), and finally disproved by Allard in 1916 (5), but in one form or another, enzyme theories were to survive in virus research for a very long time.

Hunger's spirited criticism of Woods' theories, and his dismissal of the working hypotheses of both Beijerinck and Ivanovski, were not matched by constructive proposals of his own. He assumed the pathogen to be of the nature of a toxin, a normal product of cell metabolism which under certain conditions of considerably increased metabolic activity could accumulate and cause the disturbance characteristic of mosaic disease. The toxin would then become able to enter into normal cells and induce them to form additional identical toxin. Hunger thus believed the disease could occur spontaneously, and at the same time be transmissible, a point of view expressed for fowl pox in the following decade by Sanfelice (394), just as it had been, and continued to be, occasionally claimed for rabies. Above all, it was a hypothesis which was to remain at the centre of the argument surrounding bacteriophage from 1920 onwards; in the case of bacteriophage it was strongly supported by data collected by work with lysogenic strains of bacteria (cf. Chapters 8 and 9).

Woods' enzyme theory, and Hunger's somewhat rambling musings on the nature of tobacco mosaic virus, were refuted by H. H. Allard in a series of careful experiments published between 1915 and 1918. Allard was able selectively to destroy oxidase and peroxidase activities without loss of infectivity, and vice versa, and could hence conclude that the agent responsible for tobacco mosaic disease did not belong to either of these categories, as Woods had postulated. On the theoretical side, however, Allard was unable to go along with Beijerinck's molecular flights of fancy; his conclusion was that the pathogen must be specific and particulate, in fact, 'there is every reason to believe that it is an ultramicroscopic parasite of some kind' (5).

Not long afterwards, B. M. Duggar turned his attention to tobacco mosaic virus. With Miss Karrer, later Mrs Karrer Armstrong, he devised a series of filter experiments in an attempt to determine the size of the pathogen. Their approach was comparative; they used different types of filter, and a series of colloid sols of varying particle size (e.g. casein, gelatin suspensions, lactalbumin, haemoglobin and dextrin). Their results indicated that the particle was of approximately the same size as the *particles* of a 1% solution of haemoglobin, a value they interpreted as being 30 $\mu\mu$ (1 $\mu\mu$ = 1 mμ = 1 nm). Here the authors added a footnote, referring to a paper which had come to their notice only after they had completed their study. This was the paper by Andriewsky on ultrafiltration (15; cf. Chapter 4) published in 1914. Andriewsky had similarly made comparative experiments with the virus of fowl plague, and had come to the conclusion that the particles of this pathogen were smaller than haemoglobin *molecules* for which he quoted a diameter of 2.3–2.5 $\mu\mu$. Duggar and Karrer commented: 'In discussing actual size, however, he seems to confuse the sizes of colloidal particles of haemoglobin with the sizes of molecules. Nevertheless, his conclusion is to the effect that this virus cannot be formed of cells similar to those of plants and animals at present known' (124).

The above remark reflects the uncertainties which at the time confounded any discussion concerning colloidal particles, large molecules, aggregates, etc. In particular, the concept of macromolecules, first proposed by Staudinger at this very same time in the early 1920s, was sharply opposed by a majority of his fellow chemists, who could not, or would not, let themselves conceive of organic molecules with molecular weights exceeding 5000, or containing more than 40 carbon atoms (cf. note 2, p. 28). In spite of Staudinger's own firm, and vocal, belief in his concept, it received little positive recognition until well into the 1930s, when protein chemistry began to develop with the aid of the techniques of ultracentrifugation and X-ray crystallography – once the X-ray crystallographers relinquished their firmly held belief that no molecule could be larger than the unit cell.[1]

Two years later, in 1923, Duggar presented in Philadelphia, at a meeting of the American Philosophical Society, a more comprehensive attempt to come to terms with current views on the nature of tobacco mosaic virus. The discussion was clearly sparked off by the size of 30 $\mu\mu$ for the virus which they had

[1] Olby has pointed out that to many crystallographers in the 1920s a molecule larger than the unit cell was unacceptable, and any suggestion to the contrary was treated with derision (335).

by now confirmed by repeated experiments. It compelled them to ask what could possibly be the nature of a particle of such tenuous dimensions. Duggar and Karrer Armstrong wrote:

Its life relations must be very different from those of an organism whose volume relations are to this as 37 000 to 1 or about 1,000,000 to 26. This would be the relation between the average bacterial plant pathogen and the mosaic virus. Assuming a complex organization, many theoretical questions would arise for consideration. Among these might be mentioned perhaps above all that of the surface tension conditions in such a structure, also the possibility of organization at all (membrane existence, etc.) as now comprehended. (123)

This is a representative paragraph which shows the advances made in factual knowledge and understanding since the writings of Errera and McKendrick (cf. Chapter 7), and which at the same time underlines the remaining shortcomings which were to preclude comprehension of the true nature of viruses for many years to come. Nevertheless, having set out to answer questions concerning the nature of tobacco mosaic virus, Duggar and Karrer Armstrong, with the information at their disposal, reached a very perceptive conclusion for the time at which they wrote.

They set the scene by a thorough discussion of all the evidence then available. They were at pains to refute a theory presented by Ray Nelson the previous year (327), when he had described amoeba-like bodies found in tissues of plants showing symptoms of mosaic disease. This was not a new observation; from Ivanovski (Fig. 10; Chapter 3) onwards, many authors had recorded the existence of formations of this kind, and the puzzle of their nature was a bone of contention among plant pathologists. Nelson went further than most, producing photomicrographs of the objects and identifying them as flagellates, possibly related to certain trypanosomes. He also claimed that they underwent longitudinal fission, and that healthy tissues contained no similar structures. Nelson therefore suggested that the bodies causing the disease were protozoa.

Duggar and Armstrong considered this theory at length, but their own cytological studies gave no support to Nelson's views. They also discussed the other prevailing ideas of the day, i.e. the 'enzyme' theory, the 'bacterial' theory and the 'virus' theory, and compared the effects of intensive grinding of leaves on the infectivities of tobacco mosaic virus and of spores of *Bacillus subtilis*. They summed up their findings and conclusions in a paragraph which leaves the reader with a frustrated wish to establish whether Duggar had been present in Toronto in December 1921 when Muller gave his address on variations in the gene (p. 94). They wrote:

Taking into consideration all the facts, we cannot avoid the impression, tentatively, that the causal agency in mosaic disease may be, in any particular case, a sometime product of the host cell; not a simple product such as an enzyme, but a particle of chromatin or of some structure with a definite heredity, a gene perhaps, that has, so to speak, revolted from the shackles of co-ordination, and being endowed with a capacity to reproduce itself, continues to produce disturbance and 'stimulation' in its path, but its path is only the living cell.

From this burst of activity at the beginning of the 1920s, work with tobacco mosaic virus assumed increasing importance; and gradually this plant virus emerged as a model plant virus from which could be gleaned much useful information, subsequently to be applied to viruses of vertebrates.

In the mid-1920s, Olitsky[1] brought up the question of culture of the virus when he claimed, contrary to accepted beliefs, that he had been able to grow tobacco mosaic virus in cell-free filtrates of the pulp from all vegetative parts of healthy young tomato plants (336). There was swift response from a number of plant pathologists; none was able to confirm Olitsky's results. The most immediate reply came from Maurice Mulvania in Wisconsin (318). Mulvania was a student and associate of James Johnson, who spent a lifetime studying diseases of the tobacco plant, a crop of major importance in his home state of Wisconsin. Mulvania published in 1926 a cogent study of the nature of this virus. Like Allard 10 years before, he was concerned with the chemical properties of the pathogen; working at the later date, he was able to draw on the advances made in the meantime by the rapid development of interest and knowledge in the expanding science of biochemistry.

[1] Peter K. Olitsky (1886–1964) at this time was an associate member in pathology and bacteriology at the Rockefeller Institute, with no other experience of plant virus work. His main interest was, and continued to be, neurotropic virus diseases, especially the encephalitides of man and lower animals (337).

Mulvania, backed by Johnson's knowledge of earlier literature,[1] was familiar with the papers published by Mrowka and by Andriewsky (Chapter 4) on the eve of World War I. It was unusual at this time to find reference to the work of animal virologists in papers on plant viruses, although tobacco mosaic virus had been referred to copiously in studies on animal viruses ever since Beijerinck's first paper on the *contagium vivum fluidum* (36).

Mulvania's approach was to test the effect of a number of selected treatments on the infectivity of the virus. The treatments included exposure to light (sunlight, ultraviolet light, X-rays) and heat, injection of the virus into the bloodstream of a rabbit, dialysis experiments and attempts to inactivate the virus with various bacteria. At the end of his extensive experiments, Mulvania referred to the suggestions made by Mrowka and by Andriewsky, that fowl plague virus might have the characteristics of a globulin molecule. He wrote in his conclusion: 'It is not impossible that the mosaic virus of tobacco may be of a similar nature, that is, a protein of a very simple kind, having the characteristics of an enzyme' (319).

The year 1926 coincidentally saw the announcement of Sumner's successful crystallization of urease;[2] from then on, the possibility of extracting and purifying, and maybe even crystallizing, tobacco mosaic virus by chemical means began to attract an increasing number of biochemists and plant pathologists. Certainly the first priority was purification and concentration of the virus, before attempts at chemical analysis could be made. The resourceful initiator of the virus concept, M. W. Beijerinck, had shown at the outset that alcoholic precipitation left the pathogen of tobacco mosaic disease intact (36). In 1916, Allard confirmed Beijerinck's results, with the reservation that too high concentrations of alcohol destroyed activity (5). By 1926, Walker was attempting to develop alcohol precipitation quantitatively (467), and McKinney (302), experimenting along

---

[1] Born in 1886, James Johnson had been actively working in this field from before 1910 (236).

[2] Sumner was fortuitously working with a strain of *Canavalia ensiformis* with a particularly high enzyme content; the difficulty in repeating his work with other strains did not encourage ready acceptance of his results (see 328).

similar lines, emphasized the need at this stage for standardization of the methods used. The need for reliable virus assay methods was increasingly felt at this time. Helen Purdy Beale developed serological methods for quantitative work in the late 1920s (31), while F. O. Holmes used the necrotic lesions produced in leaves of *Nicotiana glutinosa* by infection with tobacco mosaic virus (217) to develop a quantitative method reminiscent of plaque titration of bacteriophage and animal viruses (cf. Chapters 8 and 9).

In 1927, C. G. Vinson published his first results in a long series of attempts to precipitate tobacco mosaic virus out of infected juice using agents other than alcohol. He began with safranin in aqueous solution (459), and went on, with A. W. Petre, to test methodically the merits of a number of metal and ammonium salts as precipitating agents. After years of patient study, they arrived at a method based on the use of lead acetate as precipitant, with which they obtained crystals proving to possess moderate infectivity (460). The ash content was high, there was considerable loss of infectivity on re-crystallization and it was obvious that the crystals were far from being virus in a pure form. Nevertheless, work such as this held out great promise for the further application of quantitative chemical methods.

In 1931, the Western world was barely beginning to recover from the great recession, but some American institutions were unaffected, among them the Rockefeller Foundation, which opened in that year a new Division for Plant Pathology at Princeton. Its first director was L. O. Kunkel, who came to this post from the Boyce Thompson Institute, where he had been in charge of the department in which Vinson worked (254). Among his recruits in the new institute was Wendell M. Stanley who, recently returned from a postgraduate year in Germany (see 425), was instructed to apply his chemical knowledge to the problem of purification of tobacco mosaic virus. An evaluation of the existing literature led Stanley to employ the lead acetate method described by Vinson and Petre.[1] His own chemical

---

[1] Some years later, Vinson was struck down by an incurable disease. Although he lingered on until 1964, he was never to work again, spending the last 18 years of his life in the Veterans' Hospital at Kansas.

background, and the impression made by the enzyme crystallization work of Northrop,[1] seems to have left Stanley with an invincible preconception. Almost from the beginning, he modelled his approach on the methods of protein chemistry employed by Northrop; and when, in 1935, he announced the 'Isolation of a crystalline protein possessing the properties of tobacco-mosaic virus', it must have seemed to him only a confirmation of all that he had confidently expected (424).

However, Stanley's opinion was not unanimously shared by others working in this field, and some of his analytical results were viewed with unease by a number of people. Among them were F. C. Bawden and N. W. Pirie,[2] who enlisted the help of the Cambridge crystallographers J. D. Bernal and I. Fankuchen. They obtained results which differed considerably from those reported by Stanley, and they drew a different conclusion. Unlike Stanley, the British group identified their virus samples as 'liquid crystalline substances', and in addition to carbon, hydrogen and nitrogen the substance was found to contain significant amounts of sulphur and phosphorus. The latter, with carbohydrate, could be 'isolated as nucleic acid of the ribose type from protein denatured by heating...' (30).

In print for a year before the appearance of the 'liquid crystalline' paper in *Nature*, Stanley's pronouncement had great impact in the climate built up by the successful isolation of enzymes as crystalline proteins. The nucleoprotein identity of tobacco mosaic virus suggested by the work of Bawden and Pirie was all but ignored by Stanley and many others for the next couple of years. Then, gradually and apparently reluctantly, Stanley began to integrate the ideas of Bawden and Pirie with his own.[3] By 1938 he had accepted and adopted most of them,

[1] At the Rockefeller Institute, J. H. Northrop confirmed Sumner's work from 1930 onwards, when he isolated and studied pepsin in an active crystalline form (332). Sumner, Northrop and Stanley shared the 1946 Nobel Prize for chemistry for preparation of enzyme and virus proteins in pure form (328).
[2] This famous and fruitful partnership began when Bawden joined the Potato Virus Research Station, established by R. N. Salaman, at Cambridge in 1930.
[3] In 1973, Pirie summed up this interlude in his inimitable way: 'The properties that we attributed to TMV [tobacco mosaic virus] differed radically from those that Stanley had attributed to it in 1935. During the next few years he incorporated most aspects of our descriptions into his. This unanimity helped to speed virus research, but leaves unanswered the question of what it was that he isolated in 1935' (359).

and at the outbreak of World War II in 1939, a number of plant viruses had been shown to behave as nucleoproteins.

About this time, there may be found in the literature scattered examples of a return to the comparison between viruses and genes which had first been suggested by Muller in a discussion of bacteriophage back in 1921 (316; cf. Chapter 8). Astbury (see 474) and Darlington (98) drew such comparisons; but their terms of reference were not sufficiently clear at a time when the nature of genetic determination within the cell still remained uncertain. However, new evidence was being added continually. If the war years inevitably slowed down academic work in Europe, the field of filterable viruses probably suffered rather less than most other disciplines. In America, the phage group was getting into its stride undeterred; and even in Europe, work on tobacco mosaic virus continued, with some measure of information being exchanged even between groups working in countries on opposite sides in the conflict.

Throughout the war years much significant work on tobacco mosaic virus was done both in Britain and Germany; moreover, even at the height of the hostilities, they had access to each other's current papers.[1] The early collaboration with Bawden and Pirie had led Bernal and Fankuchen on to more crystallographic studies of plant virus preparations, which they found to be convenient models for 'controlled studies of colloids'. In a major paper published in 1941, they suggested a model for the composition of the tobacco mosaic virus particle, which they assumed to be built up of sub-units each of a molecular weight of 370000 (42). Two years later, Schramm in Berlin found that chemical treatment at pH values above 9 caused tobacco mosaic virus particles to decompose into sub-units of a molecular weight of 360000 (404). Schramm then joined forces with two colleagues at the Kaiser Wilhelm Institute in Berlin-Dahlem. One was G. Melchers, the other, H. Friedrich-Freksa, who in 1940 had published a thought-provoking study of the importance of forces active in chromosome conjugation for the duplication of

---

[1] The paper by Friedrich-Freksa *et al.* (discussed below) was received for publication in February 1945. It included a discussion of the paper by Bernal and Fankuchen published in the *Journal of General Physiology* in 1941 (42), and also referred to other work which had appeared in *Nature* and the *British Journal of Experimental Pathology* during the war years.

nucleoproteins (163). At about the same time, Schramm had shown that tobacco mosaic virus was virtually inactivated by enzymatic removal of its nucleic acid (404). They now proceeded together to compare parallel mutants of the virus with their parent forms, in an effort to gain further information. They were able to show that parallel mutants which induced similar symptoms in tobacco plants deviated in similar ways electrophoretically from the parental type, and that the difference in electrophoretic response corresponded to differences in amino acid composition. They also determined a molecular weight of 39 million for the intact tobacco mosaic virus molecule; with Schramm's value of 360000 for the molecular weight of the sub-units, this allowed them to conclude that the tobacco mosaic virus molecule consisted of 108 sub-units. They were also able to report that '...closer chemical analysis showed that one part of the decomposition products contains nucleic acid, whereas another one does not' (164).

Another 10 years were to elapse before it was proved that the infectivity of the tobacco mosaic virus particle is lodged in its RNA. In early 1956, the evidence was delivered almost simultaneously by Schramm, by then working with Gierer at Tübingen (178), and by Fraenkel-Conrat (160) in Stanley's recently established (1948) department on the Berkeley campus of the University of California.[1] For Schramm, it was a satisfactory conclusion to a problem he had been investigating since the beginning of the 1940s. For virology in general, it was a milestone along the way, a fitting and logical complement to the discovery, four years earlier, that only the DNA of the T-even phages penetrated into the host cell (212; cf. Chapter 9). It also marked the time when tobacco mosaic virus moved within the sphere of interest of a number of people who until then had been working in other areas of the rapidly expanding science of molecular biology and genetics. In particular, the protagonists of some of those exciting tales of the unravelling of the structure of DNA (see 244), Francis Crick, James Watson and Rosalind Franklin, turned their attention to the structural arrangement

[1] In its early days, it was described simply as a virus laboratory. It now bears Stanley's name, has a variety of departments, and much of it is given over to molecular biology in various forms. Fraenkel-Conrat is still there, as is Gunther Stent.

Fig. 23. The structure of tobacco mosaic virus, according to Franklin, Klug, Caspar and Holmes (245). (Courtesy of Dr A. Klug.)

(95) of the tobacco mosaic virus particle. Before her death in
1958, Rosalind Franklin was deep into collaboration with other
groups under Schramm at Tübingen, Caspar at Yale and
Fraenkel-Conrat at Berkeley. Rosalind Franklin was able to
show the fallacy of the current belief that the virus particle was
solid; on the contrary, it had the form of a hollow tube (161).
With Klug, Caspar and Holmes, she laid the foundations for the
famous diagram of tobacco mosaic virus which has since graced
the pages of numerous textbooks (245; Fig. 23).

Hollow tube or solid, the overall rod shape of the tobacco
mosaic virion had first been suggested by results obtained by

Takahashi and Rawlins in 1932 (436). Until then, it had been a general assumption that virus particles were spherical or near-spherical, and the determinations of particle size by ultrafiltration had been based on this assumption. The rod shape was confirmed by the studies of Bawden *et al.* in 1936 (30) and also by the first electron microscopic studies of plant viruses published by Kausche, Pfankuch and Ruska in 1939 (240; cf. Chapter 9, p. 106). Later studies have shown that although many of the known plant viruses conform to the rod shape, others are isometric. For years it was tacitly assumed that the infective material of all plant viruses, like that of the tobacco mosaic virion, was RNA; but recently workers at the University of California at Davis have found a plant virus, that of a mosaic disease of cauliflower, which is not only isometric, but which has general characteristics resembling those of certain small isometric animal viruses, such as the polyoma and papilloma viruses. In 1970, Shepherd, Bruening and Wakeman identified its nucleic acid as double-stranded DNA (408).

The plant pathologists, and in particular the field workers, have naturally enough always been preoccupied with the paths of transmission in plant virus disease. The Japanese, perennially worried about diseases affecting their rice crops, have figured prominently in this field. One Hatsuzo Hashimo, described as a rice grower, has been credited with the first experimental evidence for the role played by leaf-hoppers in the transmission of stunt disease of rice, although this cannot be documented – apparently the results were never published (see 192). Several later Japanese contributions have been available to the Western world only with difficulty, many of them being published in Japanese journals. Apart from direct mechanical transmission of virus from one plant to another by simple carriage by an insect vector, it had been known since the 1920s that in some cases prolonged incubation periods were required. Kunkel had shown this for leaf-hopper transmission of the aster-yellows virus in 1926 (253); but to obtain irrefutable proof of virus replication in the vector during the incubation period was more difficult. During the 1930s, Fukushi was able to obtain evidence of such replication which he demonstrated by transovarial passage.[1] Later (1950) Black used a similar technique to pass clover clubleaf virus through 21 generations of *Agalliopsis novella* (46) in an effort to repudiate Bawden's objections that the evidence was equivocal.

The development of an insect injection technique for direct transfer of plant virus from one insect to another by H. H. Storey working on an East African research station (431) facilitated the collection of direct evidence for the multiplication of viruses in insects, and Maramorosch and his associates used this technique to deliver final and conclusive proof of the replication of plant viruses in their insect vectors in 1952.[2]

In 1963 the finding of a complement-fixing antigen common to a plant virus and an animal virus[3] created a certain amount of stir. Unfortunately, others have been unable to confirm these results and the excitement caused by the announcement may have been premature.

For obvious reasons, much of the early work on insect transmission of plant virus disease was determined by the importance of susceptible crops to the local economy. It so happened that the rice crops of Japan and the sugar beet crops of North America were adversely affected by virus diseases carried by leaf-hoppers, and leaf-hoppers therefore figured prominently in research in Japan and in the United States. In some instances it was suspected that tobacco mosaic disease was carried by aphids, but evidence for this offered by Allard (6) and by Hoggan (216) has been queried by others who have been unable to confirm their results.[4] On the other hand, in Britain and on the European mainland pride of place was taken by the humble potato, which had been severely affected by virus disease since

[1] Fukushi used eggs laid by virus-infected females of *Nephotettix cincticeps*; nymphs hatched from these eggs on virus-free plants could transmit rice stunt virus to rice seedlings (166).

[2] Much of the meticulous and important work on plant viruses in relation to their insect vectors pioneered by Kunkel, Maramorosch and Black has been thrown into question in the last ten years by the realization that the agents of aster yellows and a number of other plant diseases are probably mycoplasma, or mycoplasma-like organisms, and not viruses (see Kenneth M. Smith (1977), *Plant viruses*, pp. 214–25. London, Chapman and Hall, 6th edn).

[3] The experiments involved strains of reovirus I, II and III (formerly ECHO (enteric cytopathic human orphan) type 10) and wound-tumour virus (433).

[4] Rosette disease of peanuts on the African continent was shown by H. H. Storey in 1928 to be carried by aphids (432).

FIG. I.

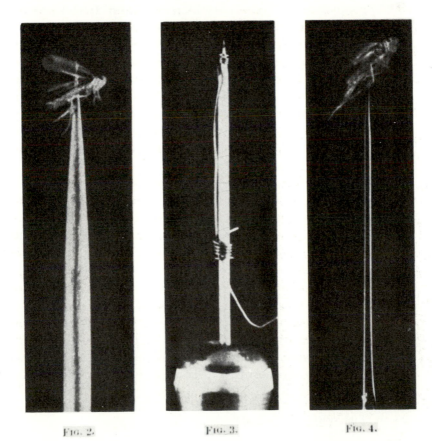

FIG. 2.   FIG. 3.   FIG. 4.

Fig. 24. H. H. Storey's photographs of the insect injection technique he described in 1933 (431).

the eighteenth century, when early outbreaks are well documented. At that time the disease was known by various evocative names, among them 'senility', 'degeneration', 'running-out' and 'crinkle' (29). Nearly two centuries later, well into the twentieth century, it was recognized that the responsible virus was transmitted by aphids, although the frequency with which plant virus diseases exist side by side in the same host caused much early confusion. An important part in the unravelling of potato virus diseases was played by the Potato Virus Research Station at Cambridge, first established as a private institution by R. N. Salaman. Here Kenneth M. Smith did pioneering work on the potato viruses he named X and Y, and here also the young F. C. Bawden was initiated into plant virus research. Being unable to select their hosts by visual means, aphids[1] rely on initial probes into the leaves on which they land. In this way they occasionally become vectors of viruses in species they otherwise do not colonize. A striking example of this mode of spread is the tulip mosaic disease, or 'breaking', a plant disease which has for centuries enjoyed the unique distinction among virus diseases of having been considered a desirable condition. Perhaps for this reason remarkably little serious research has been devoted to the aetiology and spread of this infection, while it has been amply documented in works of art and botanical illustration (cf. Chapter 1).

From research on potato viruses and their spread by insect vectors, Kenneth M. Smith proceeded to work on insect viruses proper. One insect virus disease has enjoyed almost as much and as early publicity as tulip mosaic disease, not because of its aesthetic value, but rather for its adverse effect on a provider of aesthetic values – the silkworm. The virus disease affecting the caterpillar of *Bombyx mori* was first recorded, even before the tulip was introduced into Europe, by Marco Girolamo Vida, later bishop of Alba and as such member of the council of Trent. In 1527 he published *De Bombyce*, an epic poem on the life and times of the silkworm. In a moving passage in the second book of the poem, the author of *De arte poetica* gives a vivid description

---

[1] Prominent among aphid vectors of plant diseases is *Myzus persicae* with the claim to be the vector of 120 virus diseases. K. M. Smith has referred to it as 'the arch enemy of the virus world' (see 472).

MARCI HIERONY
MI VIDAE CREMONEN
sis Bombycum Liber
secundus.

_V R S V S_ ades Nympha, inceptum iam
    perfice munus.
O _decus Italidū, fortunatißima matrū,_
Quæ faſtas feßæ Italiæ miſerata ruinas
Haud dubias pulchra ſpes nobis prole
T _u iam læta tuos natos præſtantibus auſis_   (tuliſti.
C _œlicolas ipſos ſupra ſe ferre uidebis._
A _ſpice iam quantas oſtentet corpore uires_
F _edericus puer, ut uultu decora alta parentum_
S _pondeat, u: ueniunt ſcintillæ ardentis ab ore,_
F _lagrantesꝗ; micent oculi, utꝗ; horrentia ſemper_
B _ella ſonet, pueriꝗ; agitet ſe pectore Mauors._
I _amꝗ; adeò nunc arma placent, iam feruidus acri_
G _audet equo, indomitusꝗ; animi, cupidusꝗ; pericli._
H _unc iam regna Aſiæ metuunt, oriensꝗ; ſubactus,_
G _angesꝗ;, Tigrisꝗ; tremunt, atꝗ; Indus Hydaſpes._
S _æpe hunc Euphratæ propter uaga flumina Eoæ_
B _ellantem aſpicient multa inter millia matres,_
H _orreſcentꝗ; animis hoſtem, pariterꝗ; timebunt_
E _gregio :uueni, cæcoꝗ; urentur amore,_

                    x 4   _Præſtanti_

Fig. 25. The opening page of Vida's second book on the silkworm (458). (Reproduced by courtesy of the Wellcome Trustees.)

131

of the gradual disintegration of a tribe of silkworms, by 'contagion' stealthily spreading either through 'pestilential air' or 'infected leafy boughs'.[1] Vida's description of the appearance of the diseased caterpillars of *Bombyx mori* identifies the malady as polyhedral jaundice rather than other diseases of the silkworm such as the 'pebrine' studied by Pasteur or the 'muscardine' or 'calcino' investigated by Bassi in the classic work published in 1835 (27).

More than three centuries after Vida's observations, the same disease intrigued some of the microscopists using newly developed oil-immersion lenses, because of the characteristic polyhedral inclusion bodies present in infected caterpillars. In 1883, Bolle even pointed out that the polyhedra could be identified as protein, but priority for the suggestion that the pathogen might be a filterable virus belongs to von Prowazek, archmicroscopist and author of the chlamydozoa concept (464; cf. Chapter 7). When he launched the chlamydozoa concept in 1907, the general paper had a sequel. Under the title 'Chlamydozoa II', von Prowazek presented a study of silkworm jaundice. By a combination of filtration through several thicknesses of filter paper and centrifugation of the filtrate he obtained a clear supernatant, which on microscopic examination proved entirely free from polyhedra or other particles, but which was nevertheless infective. Von Prowazek concluded that the polyhedra did not carry the virus, overlooking the fact that infection was caused by free virus particles in the filtrate *in addition* to the ones contained within the polyhedral structures.

Further contributions to the study of the aetiology of the disease were made by Acqua (4) and by Paillot (341); but the problem was not solved in detail until 1947 when Bergold, using serological tests, biophysical methods and the electron microscope, was able to show that the polyhedra consisted of a matrix of non-infective, uniform polyhedral protein in which were embedded rod-shaped particles of a DNA virus (40). These were infective when released by treatment of the polyhedra with dilute sodium carbonate solution. A year later Bergold applied his techniques to another type of caterpillar

---

[1] Vida's guess was not uninspired. The principal means of spread of the nuclear polyhedroses is by ingestion with the food plant.

disease characterized by the presence of large numbers of particles described as 'granules' ever since Paillot first drew attention to the condition in 1926 (342). In this case also Bergold confirmed the viral aetiology of the infection; the granules proved to be crystalline lattices enclosing the virus rods (41).

In 1950, Smith and Wyckoff obtained electron micrographs which suggested that not all polyhedral insect diseases were caused by rod-shaped viruses (417). Examining diseased larvae of two species of *Arctia*, they found the virus particles to be isometric rather than rod-shaped. Later it was established that these polyhedra are formed not in the cell nuclei, as are those of silkworm jaundice, but in the cytoplasm of the host cell; hence they have become known as 'cytoplasmic polyhedroses' as distinct from the 'nuclear polyhedroses'. Biochemical studies have since revealed that the rod-shaped agents of the nuclear polyhedroses and granuloses are DNA viruses, whereas the isometric virions of the cytoplasmic polyhedroses contain RNA.

More recently, yet another morphologically different insect virus has come to light. In 1954, Xeros, working with the Virus Research Unit (ARC) at Cambridge, in a routine investigation of another virus disease of the crane fly, or leatherjacket, *Tipula paludosa*, found an unrelated pathogen with unexpected characteristics (491). Crystallizing spontaneously in its living host, this pathogen causes a spectacular iridescence, and was hence christened the *Tipula* iridescent virus (TIV). Structural studies by means of double-shadowing then revealed that the virion was built as an icosahedron. Subsequently many viruses of man and the higher animals have proved to be of a similar structure. The nucleic acid of the iridescent virus was found to be DNA (416).

In 1907, von Prowazek concluded his paper on silkworm jaundice with a prophetic suggestion when he wrote:

Since, according to Bolle (1889), the much feared Nun caterpillar (*Psilura monacha* L.) is also susceptible to this jaundice, large scale infection experiments would appear to be economically very important, especially since the virus retains its infectivity when dried on to glass plates, and hence may easily lend itself to dispatch. (464)

In the 1930s a naturally-acquired virus infection so reduced the population of the European spruce sawfly in Canadian spruce forests that the damage done by this insect, which had assumed near-catastrophic proportions, became unimportant within a few years. The possibilities suggested by this chance observation were not at the time developed to any great extent, since the discovery of the chemical pesticides appeared for a while to provide the answer to the question of control of insect pests. Although a certain measure of success was obtained in attempts to control insect pests other than the sawfly with viruses, and although since then the development of resistance to pesticides as well as the observation of undesirable aspects of pesticide contamination of food, plants and animals have emphasized the need for other methods of control, the road ahead is far from clear. Bailey has pointed out that an order of magnitude comparable to the effects of epidemics considered severe or even disastrous by man would be unlikely to have sufficient impact in the case of biological control of individual insect pests (22). If in some circumstances a measure of control can be achieved by such means, even biological control may have other unlooked-for consequences, as when the eradication of one pest allows a competing species, perhaps equally undesirable, but immune to the virus used, to multiply unimpeded in its stead. It seems that a compromise between the uses of biological control and chemical pesticides may offer the most realistic hope in the battle to preserve the world's essential crops.

11. Work on the influenzas and other common virus diseases of animals and man

When the Great War came to a close in November 1918, the world was in the grip of a major epidemic which had little in common with the devastating outbreaks of smallpox or of plague which had decimated whole populations in centuries gone by. Nevertheless, its toll of misery and fatality during a period of faster travel and major troop movements was quite comparable to those earlier scourges. From the pathologist's point of view, it was a unique outbreak of lasting interest for more reasons than one, emphasizing the capriciousness of the influenza viruses and their infinite variety, as well as their peculiar affinity with diseases of other vertebrates, later to be identified as types of influenza.

At the time of the outbreak, the possibility that influenza was caused by a filterable virus was only one among several being considered. A strong contender as agent of the disease was *Haemophilus influenzae*[1] (not unreasonably, since it was frequently isolated from fatal cases), and another 15 years were to elapse before any significant progress was to be made in establishing the identity of the causal agent. In 1919 the suggestion was made that the pandemic pathogen might also be responsible for a simultaneous outbreak of an influenza-like swine disease in the United States (248). Throughout the ages, there have been many attempts to link outbreaks of infectious disease in man

---

[1] First described by Pfeiffer in 1892 (358; 397) this organism had long been relegated to its proper role of secondary invader when Shope demonstrated the intriguing relationship between swine influenza and *Haemophilus influenzae suis* infection in swine, the virus persisting in a 'masked' state in lungworms, and carried through earthworms to fresh hosts (411).

135

with simultaneous ones among his domesticated animals.[1] In the case of influenza, such inference may well turn out to have been rather more justified than in most other cases. Evidence for a connexion between swine influenza and its human counterpart of 1918–19 has been accumulating since; but whether transmission initially took place from swine to man, or vice versa, has never been clearly established.[2] In 1976 there was a warning in the United States that a swine influenza variant might again be on the rampage among human populations. The excitement created in political circles by the remote possibility of a repetition of the misery of the 'Spanish 'flu'[3] reflects alike the impact of that epidemic and the growing public confidence in the preventive powers of vaccination (although in this particular case subsequent events to some extent undermined that confidence).

With the benefit of hindsight on the other hand, it is now possible to claim that in all probability influenza viruses which are species-specific to an unrelated animal order, and an order with much less in common with the human animal than that to which the domesticated swine belongs, have had, and continue to have, a far more extensive effect on epidemiological patterns of human influenza. It was not at the time recognized as an influenza virus; and its kinship with the viruses causing influenza in man was not revealed until 1955. Nevertheless, from the earliest days of virus research, it had been a favourite object for experimental work; it was fowl plague virus.

First recognized as a filterable entity by Centanni in 1901 (p. 37), the virus of fowl plague has occupied a central place in much virus research ever since, and may with some justification be regarded as a model virus. Before he left Vienna for war duties in 1915, Robert Doerr had begun a study of this agent. After the war he resumed the work in Switzerland, and pursued it with the collaboration of a number of younger associates until

[1] This was especially true of the outbreaks of 'sweating disease' which may or may not have been types of influenza, in the fifteenth and sixteenth centuries (355; 375).

[2] Burnet and White, in *Natural history of infectious disease* (65), appear to contradict themselves in this respect on successive pages (pp. 209–10). Koen in 1919 (248) simply recorded the occurrence of an influenza-like disease in swine.

[3] The reasons for the popular prefix 'Spanish' to the 1918–19 influenza pandemic appear to be as tenuous as those for 'Dutch' in connexion with a particularly devastating disease of elm trees. The elm disease certainly did not originate in Holland; as for the influenza pandemic, its origins must remain obscure (see 89).

well into the 1930s. One of his main themes had first been revealed in a chance observation by Landsteiner and Russ (260), who in 1906 had reported that fowl plague virus appeared to be unevenly distributed within the blood of infected hens, i.e. after defibrination of virulent blood the serum was less infectious than the washed erythrocytes. For more than a decade, Doerr's fowl plague studies (116; 117) were focused on the behaviour of erythrocytes in virulent blood, and on the phenomenon of adsorption of the virus onto red blood cells. In 1941, McClelland and Hare (298) and, independently, Hirst (215) discovered the phenomenon of haemagglutination in influenza viruses. In 1955, Schäfer, at Tübingen, was able to tie up all the loose ends[1] in showing that there was a close serological relationship between human influenza A viruses and their avian counterparts as exemplified by fowl plague virus. The previous year he and his colleagues had shown that fowl plague in certain circumstances forms non-infectious, haemagglutinating particles such as observed in influenza A strains by von Magnus in 1951 (463). Von Magnus called it the formation of an 'incomplete' virus; it has since become known as the von Magnus phenomenon.

The series of papers from the Max Planck Institute at Tübingen appears to have grown from initial attempts to compare the viruses of fowl plague and of Newcastle disease.[2] The approach used by the German school in their work on the avian viruses was in a sense an extension of the outstanding studies by Leslie Hoyle in the early post-war years. Hoyle, working with strains of human influenza A viruses, examined different fractions of virus suspensions for biological activity and chemical composition. He obtained results which suggested to him that the elementary body of the virus consisted of a soluble antigen,[3] part of which was a self-replicating ribonucleoprotein, responsible, once inside the host cell, for the production of an aggre-

---

[1] Schäfer's paper was the culmination of a long series of studies on the structure and biochemistry of fowl plague virus (see 476).

[2] Newcastle disease is also a virus disease of fowl, similar in its manifestations to fowl plague, but whereas the viruses of fowl plague and human influenza are orthomyxoviruses, Newcastle disease virus is classified with the paramyxoviruses (14). Newcastle disease was described as a separate entity in 1927 (121).

[3] Soluble antigen had first been found in extracts of influenza-infected mouse lung by Hoyle and Fairbrother in 1937 (221).

gate which was ultimately completed by acquiring a lipid membrane derived from the wall of the host cell. Hoyle's definitive paper on the structure of influenza virus appeared in 1952 (220), the year when Hershey and Chase were able to show that the infective moiety of bacteriophage entering the host cell was its nucleic acid. Hoyle's ideas remained controversial for years, but have since been amply vindicated.

Until the time of Schäfer's comparative studies, for more than half a century the agents of different strains of influenza in man and the virus of fowl plague had been treated as entirely different entities, and there appeared to be no reason for comparing them. Indeed, for the greater part of the period, there was no means of comparing them, except for the results obtained by differential filtration and centrifugation. Such exercises yielded information only in the strictly limited field of relative sizes (p. 18). In 1938, Elford (139) made a comparison of values for particle diameters of various viruses obtained by different methods; for fowl plague he quoted the following values in nm: ultrafiltration; 75; microscopy and ultraviolet light photography, 75; centrifugation, 88. The comparable value for human and animal strains of influenza A, as given in modern textbooks, is 80–120 nm (allowing for the pleomorphism of this group of viruses); for fowl plague virus Schäfer quoted a figure of 70–80 nm in 1963 (399).

Progress in the area of human influenza viruses had been made in the early 1930s, above all with the demonstration in 1933 that the ferret was susceptible to the disease (419) and thus might serve as an animal model, from which it became possible gradually to adapt the virus to growth in the more convenient white laboratory mouse (13). Three years later Burnet succeeded in growing influenza virus in the developing egg (60), after several vain attempts by himself and others, following the introduction of the technique by Woodruff and Goodpasture in 1931. The original work had been concerned with the virus of fowl pox; the adaptation of the technique for use with fowl plague, and the related method which titrates fowl plague virus by quantitatively assessing its capacity to kill developing chick embryos, were long overdue. After all, Centanni had provided the foundation for such techniques in his extensive paper published back in 1902 (74), but with the exception of a brief and inconclusive paragraph on the growth of fowl plague virus in developing eggs in a French paper[1] (239) in the early 1920s, the use of eggs for the culture of viruses, even those species-specific to birds, had not been pursued for nearly 30 years. At the time of Centanni's initial experiments, the difficulties encountered with regard to contamination and large-scale incubation may well have been insurmountable; even so, it is somewhat surprising to find that so few attempts were made to make use of fowl eggs during subsequent years, when so many other potential substrates for the artificial culture of viruses were tested with such outstanding lack of success.

The isolation of influenza virus in the ferret,[2] and the subsequent advent of egg culture in this field, ushered in a period of intense activity and major advances, in London at Mill Hill where the protagonists were C. H. Andrewes, P. P. Laidlaw and Wilson Smith, and in Melbourne where F. M. Burnet led a similarly enthusiastic group. Some of these far-sighted workers were beginning to make certain comparisons with avian viruses. When Newcastle disease of the fowl was first recognized as an entity distinct from fowl plague by Doyle in 1927 (121), comparative studies of the respective viruses were made. After the discovery of haemagglutination by George Hirst, and by McClelland and Hare, in 1941, Dora Lush found that the agents of both diseases also agglutinated red cells (290). Burnet then made a comparative study of Newcastle disease virus and human influenza viruses (61); but it was not extended to include the virus of fowl plague. Thus the conclusion that fowl plague virus is in fact an avian influenza A virus, and the suggestion that it might perhaps through the processes of recombination exchange host specificity with influenza A viruses of man and contribute to the frequent occurrence of previously unencountered strains,[3] was left to Schäfer (398).

---

[1] Although the authors referred to Centanni's paper, they did not mention it in connexion with their egg culture experiments (239).

[2] Like the monkeys for Landsteiner and Popper's original polio transmission experiments (p. 50), ferrets were coincidentally available at Mill Hill in 1933 – they were used in work on dog distemper.

[3] Three separate mechanisms may account for the inconstancy of influenza viruses: (1) Antigenic drift (progressive mutative change in antigenicity); (2) Change in host specificity (swine 'flu, avian 'flu, etc.); (3) Genetic recombination. Epidemiological

Work on the development of vaccines against influenza was first undertaken with high hopes shortly after the isolation of the virus by Wilson Smith, who demonstrated the feasibility of growing influenza virus for this purpose in embryonated eggs (418) and also in tissue culture. In collaboration with C. H. Andrewes and C. H. Stuart-Harris, Smith proceeded to take part in early attempts at vaccination. The development of influenza vaccines has subsequently become a major project in vaccine centres around the world; but the variability of the antigenic characteristics of this group of viruses has placed obstacles in the path of the vaccinators, and continues to do so.[1] Nor do the influenza viruses exhibit any morphological regularity; electron microscopy in recent years has shown them to be pleomorphic, the overall shape of the virus particle being unpredictable and very different from the well-defined shape

patterns are likely to be influenced by all three, with an almost continual 'immunological drift' supplying new antigenic mutants as a result of the high mutability of the surface antigens, while more infrequently the processes of recombination between influenza viruses of human and animal origin may produce variants with greater and more sudden antigenic differences (65).

[1] It also remains to convince the public of the desirability of vaccination against colds and 'flu – in the words of Sir Christopher Andrewes: '...people don't altogether like a needle prick to give a vaccine which won't give 100 per cent protection against an infection which may not turn up anyway in any particular winter...' (11).

Fig. 26. Avian influenza A (fowl plague virus) seen in the electron microscope with the aid of two different techniques, i.e. metal-shadowing (left) and negative staining (right), demonstrating the advantage gained by the development of the latter technique. Magnification ×30000 (left) and ×180000 (right). Top: Intact fowl plague virus (strain Rostock). Centre: Ribonucleoprotein (g-antigen) prepared by ether-splitting of virus. Bottom: Haemagglutinin prepared by ether-splitting of virus. The negatively stained pictures of the two ether fractions are preparations of an influenza A virus morphologically identical to fowl plague. The metal-shadowing was done with platinum–rhodium and the electron micrographs are reproduced from Waterson, A. P., Rott, R. and Schäfer, W. (1961), The structure of fowl plague virus and virus N. *Z. Naturf.*, **16b**: 153–6. The negatively stained preparations were prepared with phosphotungstic acid, and the electron micrographs of them are by J. D. Almeida (prepared in 1969). The intact fowl plague virus was prepared by A. P. Waterson. The two ether-split fractions shown in the negatively stained preparations were supplied by L. Hoyle.

140

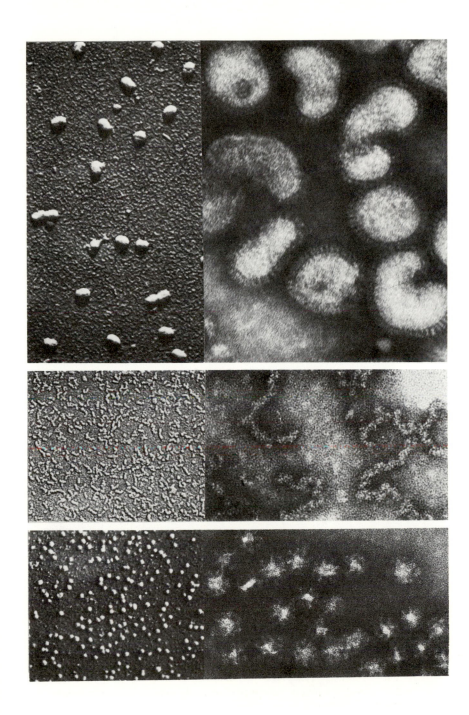

of the icosahedron, or even the elongated symmetry of tobacco mosaic virus, or the bullet shape of the rhabdoviruses.

The common cold may be less severe in its manifestations than influenza, but it is nevertheless a perennial nuisance. It was first attributed to a filterable agent by an American team under A. R. Dochez between 1926 and 1930 (112). They used chimpanzees as experimental animals. From 1931 onwards, their promising approach was taken up in Britain by C. H. Andrewes (later Sir Christopher Andrewes) who began using human volunteers, initially recruited from the student body at St Bartholomew's Hospital.[1] Later, following the end of World War II, the Common Cold Research Unit was established in the by then disused American Hospital outside Salisbury. This was a legacy from the American war effort, and the research unit owes its existence to the generosity of the American Red Cross and Harvard University Medical School, and to the initiative of Sir Christopher Andrewes. Although valuable results have been obtained there, the exercise has perhaps done more to emphasize the complexities of the situation than to solve a problem which may ultimately defy a permanent solution. Among the concrete results have been the isolation and identification of not one, but a number of viruses causing the manifestations described collectively as 'the common cold'. Most important among them are the rhinoviruses, small, isometric (diameter about 30 nm) RNA viruses (11). Later, another morphologically different type was found to produce similar symptoms; of a previously unknown morphology, they were first observed by electron microscopy in London (St Thomas's Hospital) and afterwards, for obvious reasons, named coronaviruses at The Royal Postgraduate Medical School (133). False trails were opened up with the adenoviruses, visually appealing in their almost delicate beauty of form, but with manifestations more unpleasant and severe than those of the common cold, as was also found to be true of certain ECHO (enteric cytopathic human orphan) viruses and Coxsackie viruses, all of which were at one time or other under scrutiny at Salisbury.

Among the animal viruses, the agent of fowl plague has come

[1] This and later developments are described with much amusing detail by Sir Christopher in *In pursuit of the common cold* (11).

in for a great deal of attention, as we have seen, and has been used as a model virus for reasons of convenience almost from the day it was first discovered. This is far from true of the first animal virus included among the filterable agents. Since it was first described as filterable in 1898 (p. 30), foot-and-mouth disease virus has never been claimed as a representative or convenient model virus; any publicity it has received has been of the adverse kind, as a worldwide threat to cattle industries, a ubiquitous and ineradicable menace. Friedrich Loeffler, who with Paul Frosch had headed the commission whose work first drew attention to the filterability of the agent of the disease (p. 32), worked on the problem for the remainder of his life. At first he pursued the study, focused on the development of prophylactic measures, at the Institute of Hygiene at Greifswald, on the north coast of Prussia. When repeated accidental transmissions of the disease from laboratory animals to cattle on neighbouring farms in this purely agricultural region, at a time when Germany was otherwise free of the affliction, began to cause concern, the Ministry of Agriculture reacted with commendable speed. Within a year, following a suggestion by Loeffler himself, moves were made to acquire a small island, Riems, in the bay off Greifswald. After lengthy negotiations and the necessary building works, Friedrich Loeffler could inaugurate the world's first specifically designed virological institute (albeit of a very specialized kind) on the island of Riems in October 1910.[1]

For years, the main obstacle to any real advance in the study of foot-and-mouth disease remained the lack of a suitable small (i.e. cheap and convenient) animal model. Mice, rats, rabbits and guinea pigs had been tried without success when, in 1920–21, Waldmann and Pape at the Riems institute were finally able to demonstrate that guinea pigs were susceptible after all, but that reproducible results were obtained only when infection was initiated by intradermal inoculation of the hind pad (466). Aided by this technique, Frosch became convinced, in 1924, that he had isolated the responsible organism, in the form of a small, rod-shaped bacterium (165). In the same year, a group of young

---

[1] In June 1952, the centenary of the birth of Friedrich Loeffler was marked by the re-naming of the institute the 'Friedrich-Loeffler-Institut' (377).

research workers at London's Lister Institute began a concerted attack on the agent of foot-and-mouth disease. S. P. Bedson, H. B. Maitland and Y. M. Burbury studied the immune response in guinea pigs, and the resistance of the virus to various chemicals; they also made attempts to determine the size of the agent by filtration and centrifugation experiments (this was before Elford had developed his graduated collodion membranes). Like Loeffler and Frosch a quarter of a century earlier, they concluded that the agent must be a minute, organized, living organism (34). They also made various vain attempts to cultivate the virus in different artificial media. In 1927, Maitland, at the early age of 32, was appointed to the chair of bacteriology at Manchester University.[1] It was here that he succeeded, with his wife, in obtaining proliferation of vaccinia virus by what they described as a method 'without tissue culture', in 1928 (304). Known subsequently as the 'Maitland and Maitland tissue system', or, simply, 'Maitland's medium', this later formed the basis for the preparation of the Salk vaccine against poliomyelitis, monkey kidney being substituted for the original rabbit kidney. The Maitlands then returned to the problem of foot-and-mouth disease, and in 1931 they could announce the successful cultivation of its virus through 17 passages. In this case they used guinea pig embryonic tissue cultured in heparinized guinea pig plasma and Tyrode solution.

To their immediate posterity, among which we may count ourselves,[2] it is Maitland's name in particular which is associated with the later developments in foot-and-mouth disease research, while that of Bedson is being perpetuated in the generic name of the psittacosis group of agents. When Bedson first started work on psittacosis, following the pandemic of 1929–30, little was known of the disease. It had first been described in 1880 in Switzerland, in a small outbreak confined to one household (369). The worldwide presence of the infection in parrot species which came to light after 1930 provided a classic object lesson in the biological survival of an endemic disease agent.[3] Bedson

---

[1] This early promotion served to further Maitland's career as a teacher and administrator rather more than as a research worker (305*a*).

[2] S. P. Bedson (by then Sir Samuel Bedson) died in 1969, H. B. Maitland in January 1972 and his wife 6 months later.

[3] For an excellent account of the unravelling of the psittacosis story, see *Natural history of infectious disease* (65).

and his colleagues studied the nature of the agent, and came to the conclusion that it belonged in the 'filterable virus' category; they correctly identified the 'elementary bodies' seen in the microscope as the pathogen. In a classic paper published in 1936, Bedson gave a comprehensive account of the psittacosis agent (33). Only after later developments in molecular biology and virology has it become clear that the members of the psittacosis group (which includes also the agents of lymphogranuloma venereum and trachoma) belong to a separate category,[1] characterized by the presence of *both* DNA and RNA, and replicating by binary fission. In acknowledgment of Bedson's contribution, it has been suggested that they be classified as *Bedsonia*.[2] They then share with the pleuropneumonia organism (a mycoplasma) the distinction of having spent many years in the 'filterable virus' category, only to be ousted and proved falsely included once the true nature of viruses had become known (see 18 and 134).

The mycoplasma of pleuropneumonia by its very existence was responsible for a certain amount of confusion when the concept of filterable virus was first formed (p. 33). The agent of yellow fever, on the other hand, is undeniably a virus, but confirmation of its early inclusion in the category came only after many years of sharply divided opinion and a number of tragic deaths from the disease among the people who were attempting to solve the problem. Reed and Carroll, in their classic work at the turn of the century (p. 47), had shown that a filterable virus rather than Sanarelli's *Bacillus icteroides* was the agent responsible for yellow fever. But throughout the 1920s, Hideyo Noguchi made determined attempts to show that yellow fever was caused by a spirochaete which he called *Leptospira icteroides*. Between 1925 and 1927, work in West Africa solved the problem of the aetiology of yellow fever. Stokes, Bauer and Hudson (429) finally confirmed that a virus was the causative agent; but by the time the results were published, the disease had claimed both the victorious Stokes and the defeated Noguchi.[3]

[1] They have disappeared from *Viruses of vertebrates* (14) between the first (1964) and second (1967) editions.

[2] Otherwise they are known as *Chlamydiae*.

[3] Noguchi's colourful personality has evoked comment from friend and foe. An objective and well-balanced account of his life and work has been given by P. F. Clark (82).

When Reed and Carroll showed yellow fever virus to be filterable, it was the first virus pathogenic to man to be included in the new category. Since at the same time they confirmed Finlay's suggestion that it was transmitted by mosquitoes, it was also the first member of what was later to become known as arboviruses, i.e. the arthropod-borne viruses (14). In 1908, Doerr filtered the agent of the phlebotomus-borne pappataci fever which, with dengue, rates as a comparatively mild nuisance compared to the many lethal encephalitis viruses of man and animals which are also carried by arthropod vectors, among them Japanese encephalitis, Murray Valley encephalitis, St Louis encephalitis and the Eastern and Western equine encephalitides (14).

Many arboviruses have been discovered since 1927, in the course of work on yellow fever and 'jungle' yellow fever in Africa and South America; but some viruses transmitted by arthropods must, for morphological and other reasons, be included in different groups. This applied to the virus of rabbit myxomatosis, which in electron micrographs is very similar to vaccinia virus, and which with the viruses of Shope fibroma and of Californian rabbit fibroma belongs to the poxvirus group. The virus of myxomatosis has taught man a number of lessons concerning the results of interference with his naturally-balanced ecosystem;[1] the fibroma viruses occupy a place in the history of tumour virology (Chapter 13).

[1] The numerous lessons to be learnt from the study of this disease and its virus are discussed by Fenner and Ratcliffe in *Myxomatosis* (152).

12. The 'eruptive fevers'

With the exception of such atypical outbreaks as the 1918–1919 pandemic, influenza is rarely a killer but continually and perennially a nuisance to man. Its frequent changes in antigenicity make the hope of lasting immunity, acquired either naturally or through vaccination, a vain one. The situation is very different with regard to a morphologically related virus, that of measles. Like the influenza viruses, the pathogen of measles is an RNA virus with helical symmetry and a lipoprotein envelope; but clinically and historically it has closer links with the disparate group of viruses causing the eruptive fevers associated with childhood. Making regular forays into young, previously unexposed populations, they produce lasting immunity and few complications in healthy children who encounter them at the 'optimum age' between six and twelve years.

Similar to influenza in their initial manifestations, they are all characterized in later stages by more or less severe rashes on the face, trunk and extremities. In typical cases the rashes are characteristic of each disease and sufficiently different to allow differential diagnosis, although for the historian of epidemiology the inadequacy of early descriptions often makes identification uncertain in retrospect. To the difficulties caused by inaccurate descriptions in the early literature is added the confusion created by a vague and indefinite terminology. Since the writings of Rhazes (367), measles has been known as a separate entity, but was often regarded, perhaps also by Rhazes, as a milder form of smallpox rather than a completely different infection. Since the concept of infection at the time was anything but clear-cut, there is little point in attempting to analyse Rhazes' precise terms of reference. The same may be said of the subsequent literature until well into the eighteenth century. Then, in the age of enlightenment, physicians (among them Thomas Fuller (1730, 167) and William Heberden (1767, 202)) began to draw clearer distinctions, and slowly and gradually there emerged a composite picture of a number of eruptive fevers. Again, smallpox was the protagonist which helped to focus interest on its lesser brethren,[1] and the severe epidemics in Britain and on the European mainland throughout the eighteenth century[2] are reflected in the copious literature on smallpox and on inoculation practices. In 1767, William Heberden the Elder gave the first differential description of chicken-pox (202), while rubella was recognized as a distinct entity in the early nineteenth century. Maton, in 1815, singled it out for the slightness of its symptoms (310) in what had been thought to be an outbreak of scarlatina in one of 'our great public schools'. In Germany, where it was known as Rötheln, credit for the differential description goes to Wagner, with a paper written in 1829 (see 311). The German name was in general use until Veale, in 1866, described it as 'harsh and foreign to our ears' (456) and suggested its replacement in the English literature by rubella.[3]

The widely disseminated interest in smallpox inoculation in the eighteenth century produced as a corollary attempts to prevent in a similar way the more severe manifestations and complications seen in the measles. In 1757, Alexander Monro (secundus) suggested that such a procedure might be tried; the following year Francis Home, a fellow Scot, put Monro's theory into practice (whether independently or prompted by Monro's suggestion is not clear, see 206). It seems unlikely that Home's inoculation experiments were as successful as he claimed, but they were well received by his contemporaries, and were quoted (though anonymously) by George Baker in *An inquiry into the merits of a method of inoculating the small-pox, which is now practised in several counties of England* (23), a work which was otherwise mainly concerned with a refutation of the dangerous practice

---

[1] Although 'brethren' is probably too strong a term; later techniques have shown that there is neither a morphological nor serological justification for a common grouping in spite of the common manifestation of some kind of eruption.

[2] For details see Charles Creighton's *History of epidemics in Britain*, 1894 (93).

[3] However, there was little general recognition of rubella's separate identity until it had been fully discussed at the International Medical Congress in London in 1881.

of 'preparing' patients for inoculation by a rigorous regimen of starvation and general discomfort.[1]

Jenner's introduction of cow pox vaccination on the threshold of the nineteenth century, and the resulting controversy which long outlasted Jenner himself and continued through the century, tended to overshadow other work on the eruptive fevers. But in 1846–47 the attention of the medical world, poised on the brink of great discoveries in the field of infectious disease, was claimed by P. L. Panum's classic study of a measles epidemic in the Faeroes (343). By virtue of their geographical isolation, these islands were a perfect area for a controlled study of the course of an epidemic. A milestone in the history of epidemiological studies, and a model for many later ones, Panum's work paid little attention to the nature of the aetiological agent. In the event, another century was to elapse before any real advance was made with regard to identification of the various viruses responsible for individual eruptive fevers. Then, almost simultaneously, their morphology as revealed by the electron microscope, and the differences found in nucleic acid composition, demonstrated once and for all how very different were the individual pathogens severally responsible for those diseases characterized by rashes and pustules.

In the case of measles, the filterability of its pathogen had been confirmed in 1911 by J. F. Anderson and J. Goldberger (7); but little was added to this limited knowledge until after the perfection of tissue culture techniques by Enders and his colleagues in the late 1940s (p. 76). Further elucidation was directly derived from observations of cytopathogenic effects made possible by the studies of Enders and his team. Fittingly, it was achieved by a young associate of Enders' working closely with Enders himself in 1954 (144). Within another decade, electron microscopy and protein chemistry provided detailed information concerning the structure and nucleic acid content of the measles virion.

The mildness of the clinical manifestations of rubella meant that for a long time the disease received very much less attention

---

[1] Such misguided attempts to 'prepare' the patient, which often had disastrous consequences, may owe something to a misinterpretation of Sydenham's seventeenth-century belief in 'constitution' as a factor of paramount importance in pathogenesis.

than the others mentioned in this chapter. On the other hand, there appeared to be no valid reason for objection to producing experimental rubella in human volunteers, and from 1938 onwards a number of studies transmitting the disease to monkeys and to human volunteers by means of cell-free filtrates suggested a viral aetiology. This was finally confirmed in the early 1960s, again as a direct result of the application of the new tissue culture techniques and the observation of cytopathic effects. Shortly afterwards, electron microscopic studies of the morphology of rubella virus began to appear: but by then, rubella virus had for long been known to be very much more of a menace than might have been suggested by its mild clinical manifestations.

In 1941, there had appeared in the *Transactions of the Ophthalmological Society of Australia* a paper with the title 'Congenital cataract following German measles in the mother' (181). Because of the prevailing war, and because the journal was not one to be read as a matter of course by clinical microbiologists and paediatricians, its significance was only slowly realized. By the time it became possible to isolate the virus in tissue culture, the severity of congenital malformations and the need for the development of a vaccine had become abundantly clear (p. 6). Since Gregg's observations in 1941, other viruses have been indicted as potential perpetrators of congenital malformation, among them the cytomegaloviruses, and it is unlikely that we have yet seen the end of that particular chapter of the history of virology.

As its name indicates, the manifestations of chickenpox can at first resemble a mild case of smallpox, but it would be wrong to conclude that there might be a similar relationship between the aetiological agents. However, whereas it cannot claim any connexion with the poxvirus group, the virus of chickenpox, or varicella, does belong to another major group of viral pathogens, i.e. the herpesviruses. A possible connection between varicella and herpes zoster was first mooted in a paper by von Bokay in Vienna in 1909 (461); three years earlier Tyzzer had described nuclear inclusions from skin lesions in varicella which were later to prove typical of the herpesvirus group (454). In 1919, Paschen, extending his interest in smallpox and vaccinia (Paschen bodies) to include chickenpox, suggested a viral aetiology for varicella. Paschen supported his claim with findings in the light microscope, findings which were later amply vindicated in the age of electron microscopy (346).

The revelation that the viruses of varicella and of herpes zoster were mutually indistinguishable added yet another element of complexity to the many-faceted saga of the herpesviruses (some of them will be found in the following chapter, in a very different and more sinister context). The term 'herpes' has been part and parcel of medical literature since before the time of Hippocrates. Its meaning has changed greatly over the centuries, and there is no evidence that 'herpes' at the time of Hippocrates necessarily referred to any of the conditions which today bear the name.[1] Derived from the Greek verb 'to creep', it appears originally to have been used of ulcerative cutaneous lesions in skin cancer, lupus vulgaris, erysipelas, ringworm, eczema and perhaps even smallpox. Such catholicism of expression does not facilitate the task of the historian, but there has been no dearth of attempts to identify the various conditions to which the term has been applied in the Hippocratic writings, by Galen, Celsus, Avicenna and many later writers. Not until the eighteenth century did some nosological order begin to emerge. According to Beswick (43b) Daniel Turner gave the first adequate description of herpes simplex under that name in 1714; but he also pointed out that although Turner included, correctly, zoster as a species of herpes, he still considered ringworm another species of herpes, as did many subsequent writers until the middle of the nineteenth century. In 1730, Thomas Fuller singled out 'herpes miliaris' as one of the exanthematous conditions not to be confused with smallpox; his brief description sounds more like herpes simplex than zoster which according to Turner had been the earlier sense of 'herpes miliaris'. In the early nineteenth century, Bateman, who had inherited some of his material from Willan, distinguished between six different kinds of herpes, all characterized by one or more localized crops of small superficial vesicles healing spontan-

---

[1] The history of the term 'herpes' has been fully discussed in a paper by T. S. L. Beswick in 1962 (*Med. Hist.*, **6**, 214–32).

eously after 10–12 days. He also stated that these conditions were not infectious.

However, any real attempts to determine the aetiology of conditions of the herpes group belong strictly to the twentieth century. In Prague before the outbreak of World War I, Wilhelm Grüter had been studying vaccinia lesions on the rabbit cornea with material from human keratitis and from vesicles of herpes labialis (185). His results were confirmed by Löwenstein who in 1919 (280), in the course of his experiments, suspected 'the possibility, even the probability, that the virus of herpes may belong to so-called invisible, filterable viruses'. Filtering suspensions of herpesviruses through Berkefeld filters he obtained negative results, but commented that this was not necessarily the end of the matter since 'many early attempts to filter vaccinia virus had also been unsuccessful'.[1] Shortly afterwards, Lipschütz published his classic paper 'Untersuchungen über die Ätiologie der Krankheiten der Herpesgruppe (Herpes zoster, Herpes genitalis, Herpes febrilis)' in which he identified the characteristic inclusion bodies which he named '*Zosterkörperchen*'. He also postulated that the agent causing herpes febrilis and herpes labialis was different from that of herpes progenitalis, a view which was not vindicated until long after Lipschütz's death when refined techniques could justify the division of the herpes simplex agent into two sub-species according to immunological specificity and base ratios of DNA (as well as their site of infection) (272).

At about the same time, Robert Doerr reported the occurrence of encephalitic manifestations in rabbits with severe herpetic cornea lesions (115), and Levaditi and Harnier isolated a pathogen with herpesvirus characteristics from a patient who had died of von Economo's encephalitis.[2] In 1920, Levaditi concluded that the virus of herpes might be a less virulent form of 'the virus of encephalitis'. He continued with his studies of inclusion bodies which he identified as protozoa, as in '*Encephalitozoon cuniculi*' and, in a study of Negri bodies, '*Encephalitozoon rabiei*' (269).

[1] The vaccinia virus particle measures approximately 240×300 nm, while the size of the herpesvirus particle is about 120×220 nm.
[2] First described by von Economo in 1917 (462), this epidemic encephalitis was to present problems for years to come.

The 'eruptive fevers'

In June 1930 the *Lancet* published a paper by Andrewes and Carmichael (12). Its modest title 'A note on the presence of antibodies to herpesvirus in post-encephalitic and other human sera', gave scant indication of the far-reaching implications of its findings; for Andrewes and Carmichael had found that a large proportion of the normal adult population carry neutralizing antibodies against herpesvirus in their blood, and that recurrent herpes occurs only in those with such neutralizing antibodies. This was contrary to all the accepted tenets of current immunology, and created an aura of confusion around the herpesviruses in particular, at a time when the concept of filterable viruses was in any case on uncertain ground. Two years previously, in 1928, Boycott had written:

I am inclined to think that [viruses] are both the cause and the result of their diseases as Sanfelice suggested for *epithelioma contagiosum*. [1] Somehow or other a virus arises in an animal or plant and by its action on the tissues causes them to produce more of itself. Some viruses (e.g. smallpox) acquire a considerable capacity of spreading from infected to normal individuals and the majority of cases of the disease are so caused; the virus is on the way towards independence. Others (e.g. herpes) have little or no power of dispersion and most cases are due to the virus arising *de novo* under the appropriate stimulus (whatever that may be). (54)

Ten years later, and three years after the crystallization of tobacco mosaic virus, Doerr could still offer the opinion that '...the agent of *herpes simplex* is not a pathogen maintained by a chain of infection; on the contrary, it is formed endogenously in the human organism' (118).

It is no longer believed that encephalitis lethargica, or von Economo's encephalitis, is caused by a herpesvirus. But in 1934, a laboratory worker in the United States was bitten by a seemingly healthy Macacus rhesus monkey. Eighteen days later he died of acute ascending myelitis (390). The pathogen was found to be a virus related to those of herpes simplex and pseudorabies. In monkeys it produces merely a mild, herpes-like infection, but when transmitted to man the change in host specificity causes a conspicuous increase in virulence, and nearly all cases since the one originally reported by Sabin and Wright have been fatal.

[1] Sanfelice had first propounded this theory in 1914 (394) and elaborated on it in 1927 (395); cf. p. 46.

This chapter would hardly be complete without a return to the great protagonist which turned attention to all the other 'eruptive fevers'. Although considered by authors from Rhazes to Fracastoro to be also a disease associated primarily with childhood (p. 5), smallpox has later shown a pattern of a different kind. In the end, the virus proved even morphologically to be in a class all of its own, unrelated to other filterable DNA viruses outside the pox group, with no apparent symmetry, either icosahedral or helical. Its position at the head of any table recording sizes of viruses accounts for the fact that the elementary bodies were seen, albeit just, in the light microscope even before the formation of the concept of filterable viruses (p. 21). It also explains why the poxviruses were added to the growing list of these viruses only at a relatively late stage, in 1905 (p. 49), after many vain attempts at filtration.

Keber's filtration experiments (241), and Buist's staining of vaccinia lymph (56) in the second half of the nineteenth century were cogent studies, as were Chauveau's series of papers dated 1868 (78), at a time when more attention was lavished on the often acrimonious debate for and against smallpox vaccination. Prominent among those who vociferously objected to vaccination until the end of the century was Creighton (93). An author less frequently referred to in medical texts wrote warmly in defence of the procedure, in a series of letters to newspapers. He signed himself Lewis Carroll.

While progress was slow in the attempts to gain information concerning the aetiology of smallpox, there were nevertheless interesting developments[1] which brought into focus an aspect of some virus diseases which throughout the early years of virus research contributed more to the general confusion than to any solution of questions of aetiology. Guarnieri discovered the inclusion bodies named after him in 1892; he regarded them as stages in a protozoan life cycle, and named the organisms *Citoryctes vaccinae* and *Citoryctes variolae* (186). Rivolta in 1869 (374) and Bollinger four years later (49) made similar observations for fowl pox (epithelioma contagiosum). In 1905, von Prowazek re-introduced the term 'elementary bodies' first used by

---

[1] Not surprisingly, the bulk of experimental work when possible has been performed with vaccinia rather than variola virus.

Chauveau[1] in 1868 to designate the minute, just visible bodies seen inside the inclusions and suspected, correctly as it turned out, of being the infective particles. Also in 1905, Negri finally succeeded in passing the pathogen through filters. Twenty years later the methods developed by Barnard in ultraviolet microscopy (see 139), and by Elford and by Bechhold in the area of ultrafiltration and ultracentrifugation (p. 18) brought results which suggested that the viruses of the pox group were among the largest particles in the filterable virus category. In 1948, electron microscopy came into its own in diagnosis of poxviruses in studies (379) demonstrating at the same time that the virus of varicella had little in common morphologically with the poxvirus group. Improvement in technique allowed observation of the finer structure, and so-called negative staining with phosphotungstic acid revealed the presence of sub-units on the particle surface (cf. p. 141). These structural studies appeared in the early 1960s, when it also became clear that the nucleic acid of the poxviruses is DNA (237).

Thus a very great deal of detailed information has been gained about the virus particles of vaccinia in particular, and of poxviruses in general. Nonetheless, a question remains which now, on the eve of eradication of smallpox, seems unlikely ever to be answered. It concerns the origin and true identity of vaccinia virus. Studies by Bedson and Dumbell have made it not unreasonable to assume that it might be a hybrid of smallpox and cow pox viruses. Mixed infections which could have favoured such hybridization, through recombination, certainly occurred from the very early days of Jenner's efforts to popularize vaccination, as in the famous episode at the London Smallpox Hospital when Baron, Jenner's faithful friend and biographer, described the two viruses as having become 'commingled' (26). That was in 1799; the possibility of unravelling the identity of vaccinia virus has become increasingly remote ever since. Not only did vaccine from that original strain subsequently form part of the material which was widely distributed both in Great Britain and on the European mainland, but in some areas it has subsequently become accepted practice to mix

[1] Chauveau called them '*granulations élémentaires*'. His terms of reference were necessarily rather more vague than von Prowazek's (78; 465).

vaccine strains. In Germany, virulent variola virus from human sources has in some cases been added to vaccinia batches thought to be in need of boosting. With such practices still accepted in the twentieth century, there can be little hope of arriving at a satisfactory answer to the question of the origin of vaccinia virus on the basis of the necessarily limited, and not always objective and reliable, historical evidence available from the early nineteenth century.

13. Tumour virology

For centuries man has been preoccupied with proliferative lesions of his own and other species. The grotesque appearance of some tumours was early reflected in medical illustration and in that curious borderline area between sculpture and medical illustration, the medical wax sculpture (39; 326).

In 1739, Lorenz Heister was making the assumption that mammary tumours were, or at least might be, transmissible when he described the precautions to be taken with 'infected blood' during mastectomy (p. 6). A century later, French interest in physiological animal experimentation was at its height, fostered by men such as Magendie and Paul Bert. By the middle of the century, rapidly improving microscopes and associated techniques encouraged development in a number of areas and led to great advances in histology and pathology, as well as the rise of bacteriology (cf. p. 7). These thriving disciplines came together in an extensive treatise on chest disorders published by Velpeau in 1854. Velpeau was above all a gynaecologist, and in his discussion of mammary cancers he included a chapter on the aetiology of tumours, with a special section devoted to 'contagion'. Here he cited a number of authors, as well as experiments of his own, in support of a theory of transmissibility of certain tumours (457). The evidence presented is not impressive, consisting for the most part of single experiments. Langenbeck may have been the first (1840) to perform an experiment in which material from a human tumour was injected into the femoral vein of a dog. At autopsy the dog was found to have two to three small round tumours the size of lentils on each lung, in addition to one larger tumour on the right lung.

Goujon (1866–67) performed similar experiments, sometimes

with positive results, sometimes not. He also implanted sections of tumours under the epithelium of rats, guinea pigs, etc. and observed the formation of small tumours adjacent to the graft as well as in the viscera. At the same time, Wilhelm Müller, in Germany, suggested that something was transmitted from tumours of the thyroid to induce metastases in other parts of the body such as lungs and lymphatic glands (317).

Even those who believed that tumours developed in experimental animals as a direct result of the injection of human cancerous material offered little in the way of theories concerning the mode of transmission. Then, in the last years of the nineteenth century, the first filtration experiments with tobacco mosaic virus and with foot-and-mouth disease virus created a flurry of excitement in the world of pathology. When very shortly afterwards Sanarelli described myxomatosis and pronounced its agent 'invisible', some of those involved in cancer research took notice. Today myxomatosis is included in the poxvirus group, and the lesions are not regarded as virus-induced tumours; but in the general climate prevailing in pathology in 1898, Sanarelli's paper may have inspired Leo Loeb to an attempt to transmit cancer of cattle by means of a cell-free filtrate (see 218). Like other such attempts at the time, it failed. Nevertheless, the idea of an aetiological role for filterable viruses in some forms of cancer was to persist and grow in strength, albeit slowly.

At the Institut Pasteur in Paris, the young Amédée Borrel was one of the team that demonstrated the filterability of the agent of bovine pleuropneumonia. Although it was a discovery which hindered rather than helped the understanding of the virus concept in years to come, it did lead Borrel on to the study of other virus diseases, and in particular those he called infectious epithelioses in which the virus invades the epithelial cells and causes proliferation. Ten years earlier, Borrel had written his thesis on epithelioma; inevitably, he was drawn to make comparisons. In a classic paper in 1903, he declared of the epithelioses and epithelioma: '*Il'y a analogie, il n'y a pas identité*' (53). He was to devote the rest of his life to attempts to determine the role of viruses in the pathogenesis of cancer.[1]

[1] The contributions of Borrel in this area have been discussed in a paper by Le Guyon (265).

In 1910–11 came the first experimental evidence which could be taken as corroboration for his theories. In that year Peyton Rous was able to demonstrate that a tumour-producing agent could indeed be transmitted with a cell-free filtrate in series in chickens (381). There was far from general approbation of Rous' results and conclusions. They were not so much questioned as totally rejected by the majority of his contemporaries; most vehemently of all were they denounced by certain schools in the forefront of cancer research.[1] Rous himself never wavered in his belief in the significance of his results[2] for which there had indeed been some indirect support in a study published by two Danish pathologists in 1908 (140). They had isolated a filterable agent inducing fatal leukaemia in chickens; but quite apart from the negative attitude to the concept of infectivity in cancer, leukaemia at that time was not regarded as a form of cancer. Consequently there was no reason for drawing analogies between Rous' results and the observations made by Ellerman and Bang.

In spite of the opposition, Rous continued undeterred and unconcerned his studies of chicken sarcomas, describing subsequently other types of 'filterable tumours' of the hen (382). But for many years, the writings of Borrel and Rous' studies of chicken sarcomas remained as voices crying in the wilderness. In the early 1920s W. E. Gye began in London a study of Rous' chicken sarcoma which convinced him that 'the central problem of cancer' would ultimately prove to be one of a specific virus-induced disease.[3]

While such theories remained anathema to the majority of those who had 'devoted themselves to cancer research' (Gye, 188), the attitude of some of Rous' colleagues at the Rockefeller Institute was more positive. T. M. Rivers, in a study of pathological changes in rabbit myxomatosis infections, compared the myxoma to Rous' sarcoma and suggested that the myxoma

---

[1] This was especially true of the powerful German school, although Rous himself was apparently fond of recounting the remark by a visiting British oncologist: 'But my dear fellow, don't you see, this can't be cancer because you know its cause' (382; 413).

[2] Fortunately, he lived long enough to see his work vindicated, and fully recognized, dying at the age of 90 replete with honours.

[3] With W. J. Purdy, Gye later turned to an immunological approach, and they eventually published *The cause of cancer* in 1931 (190). Their results and conclusions have been criticized, but many of Gye's theoretical arguments, if not his experimental results, have later been vindicated (189).

might 'bridge the gap' between Rous' tumour and other virus diseases (371). This reasoning was taken a step further by R. E. Shope when in 1931 a freshly shot rabbit with tumour-like growths was examined at the Rockefeller Institute. Shope was able to show that the condition was transmissible in rabbits by means of an agent filterable through Berkefeld V and N filters (p. 17). The virus was found to be related immunologically to that of myxomatosis,[1] although clinically and pathologically the conditions were entirely different (409).

In the following year, Shope's attention was drawn to a papilloma occurring in the wild rabbit population of the states of Iowa and Kansas. It had long been known to trappers in the area, and Shope was now able to demonstrate that this also was caused by a filterable virus, in this case unrelated to either myxoma virus or the virus of the previously described infectious fibroma. Although the new virus was transmissible in series through wild rabbits and could be transmitted to domestic rabbits, it was not transmissible in series through the latter, but seemed to enter into a 'masked' or 'occult' state in the foreign host (410). When Rous and Beard subsequently studied the papillomas induced in domestic rabbits, they discovered another characteristic. The benign papillomas induced in domestic rabbits with the virus from the wild ones gradually developed into malignant carcinomas (383).

While Rous, Borrel, Gye and their followers saw the work on transmissible animal tumours as corroboration for their theory that cancer was caused by viruses, others interpreted the results differently and used Rous' sarcoma virus as an argument in favour of the theory that many, if not all, viruses arose *de novo* in the cells of affected tissues. The way in which the same results can be claimed to support diametrically opposed views may be illustrated by two sentences written in Britain within the short span of three years. In 1925, Gye, concluding a paper on Rous' sarcoma virus, wrote: 'These researches have led me to look upon cancer – using the term in its widest sense – as a

---

[1] On injection into domestic rabbits the filtered tumour suspension produced large tumours which regressed spontaneously after 10 to 12 days. The animals were subsequently found to be immune to not only this virus, but also that of myxoma.

specific disease caused by a virus (or group of viruses)'[1] (188). Whereas Boycott, in 1928, decided that, 'As to its origin, all the evidence seems to concur in indicating that the Rous virus arises *de novo* in each tumour. There is no epidemiological evidence that cancer comes into the body from the outside; everything we know supports the classical view that it is a local autochthonous disease' (54).

The suggestion that hereditary factors might influence the incidence of mammary cancer in mice was first made in work by J. A. Murray in 1911 (322). For many years, progress was slow and limited to breeding experiments, and the tumours were considered to be of unknown aetiology. Then, in 1936, J. J. Bittner had the idea of letting newborn mice of a high-tumour strain be nursed by adult females of a low-tumour strain. Working initially with a relatively small number of animals, Bittner soon extended his work and could confidently claim the existence of a 'milk factor' influencing the incidence of this type of cancer. The agent transmitted with the milk was not expressed immediately, and the affected mice remained apparently normal until they reached middle age, when they would develop mammary malignancies (45). Bittner's results demonstrated very clearly the need for serious re-thinking of some basic assumptions. What had looked like spontaneous neoplasms obviously had a viral aetiology; on the other hand, there were evidently other factors involved as well, possibly of a hormonal nature.

Further considerable steps forward came in the 1950s, when work on the mouse leukaemia viruses started. There proved to be several of these, and they can be placed in two groups, one which comprises the mouse leukaemia viruses of the 'thymic' type, described by Gross, Graffi and Moloney, while the other, 'splenic' types have been studied by Friend and Rauscher. The first to be described, in 1951, was Gross' virus which is vertically[2]

---

[1] This paper was prefaced by an editorial note to the effect that, with its accompanying paper by J. E. Barnard on ultraviolet microscopy of the virus, it offered 'detailed description of a prolonged and intensive research into the origin of malignant new growths, and they may present a solution of the central problem of cancer' – the latter perhaps a somewhat extravagant claim.

[2] Gross was able to demonstrate vertical transmission through the *ovum* from one generation to the next for one or two generations, but not indefinitely.

transmitted in certain inbred strains of mice (184). As was the case with Bittner's virus, the pathogen described by Gross did not become clinically manifest in the infected young until they reached a certain age, and presumably a particular hormonal constitution. Two years after his initial observations, Gross discovered that he was in fact studying not one, but two individual viruses, and that in addition to leukaemia some of the inoculated mice developed carcinomas of the parotid glands caused by a second virus. Gross called the second agent parotid tumour virus. This was soon to attract attention of its own through the work of Stewart and Eddy who in 1957 discovered that growing the parotid tumour virus in monkey or mouse embryo tissue culture caused a marked increase in its virulence. It then became capable of inducing tumours of a number of different types in a number of hosts such as rats, hamsters and rabbits. They re-named the agent polyoma virus (427), and as such it has been extensively studied as a valuable tool in oncological research. Not long afterwards Eddy and her colleagues turned their attention to a simian virus, SV40. The agent of an apparently harmless, latent infection in the kidney cells of certain species of monkey, this virus was found to induce neoplastic disease in baby hamsters (131).

Thus, 50 years after Rous' initial experiments with the filterable virus of chicken sarcoma, the picture had become very much more complex, but also in some respects clearer. The study of murine leukaemia viruses and of SV40 suggested that some viruses may be either latent or even silent agents in their natural hosts, only to show strongly oncogenic properties when introduced into certain other animal species. In fact, several types of adenovirus which cause respiratory infections in man but, so far as is known, are not oncogenic in this natural host, can initiate tumours in hamsters and rats (449). Furthermore, observations on these viruses suggest that they behave in their foreign hosts much as the papilloma virus of the wild rabbit had been shown to behave in domestic rabbits more than 20 years earlier: they pass into a 'masked' state.

The technique developed by Gierer and Schramm for their demonstration in 1956 that the infectivity of the tobacco mosaic virus was lodged in its nucleic acid component, was brought into use in the field of tumour virology in 1960. It enabled Ito to show with reasonable certainty that the tumour-producing factor in papilloma virus was likewise to be found in its nucleic acid, probably DNA (226), and also opened the way for Shope to present a tentative explanation of the 'masking' phenomenon he had observed in rabbit papilloma virus 25 years earlier. Such behaviour could be explained, according to Shope, by an assumption that the virus exists in the wild rabbit papillomas in two forms, one of which is the complete virus consisting of nucleic acid plus protein, while the other, immature version has not acquired a protein coating but remains naked viral nucleic acid. The different compositions of the two forms would then explain their different activities. The mature, complete virus would be necessary to induce the formation of tumours, whereas the incomplete (naked nucleic acid) form could maintain neoplasia, but not pass on the infection (412).

Although many points remain obscure, the viruses discussed above are freely described as 'oncogenic' by those studying them. A far more cautious terminology is employed in the description of another group of viruses which has much more recently become a focus for attention in tumour virology. They are the tumour-associated herpesviruses which have been found, as the name suggests, in association with tumours in man and in a number of animals. Their possible role in the initiation of the tumours with which they are associated is far from clear; but then the conditions are, like herpes simplex, of complex aetiology. One extensively studied disease of fowl falls into this category. Marek's disease was first described in 1907 (307) when it appeared to be another poultry disease characterized, like fowl plague, by involvement of the nervous system. In 1929, at a time when the disease was becoming a problem to poultry farmers in Europe and the United States, Pappenheimer and his colleagues found a high incidence of lymphoid tumours of the ovary in affected chickens. As poultry farming became increasingly intensive following World War II, the disease appeared to gain in virulence, and to change from the classic type characterized primarily by neural involvement to an acute form in which an important manifestation is the development of visceral lymphoid tumours.

Isolation of a cell-associated herpesvirus from cases of
Marek's disease was finally achieved in 1967, when Churchill and
Biggs were able to grow a cytopathogenic agent in tissue culture
(79). The cytopathogenic effects were similar to those obtained
with certain herpesviruses, especially the varicella–zoster agent,
and electron micrographs supported the cytopathogenic findings. The results brought Marek's disease into line with two
other conditions in which herpesviruses had been found in
association with tumours: Lucké's renal carcinoma of the
leopard frog, and, more important to the medical sciences,
Burkitt's lymphoma of man. But two years after the virus
isolation, Churchill and his colleagues achieved for the Marek
virus a more direct indication of involvement in the aetiology
of the disease than can so far be claimed for the other two
conditions. They were able to protect chickens against challenge
with virulent virus by means of a vaccine prepared with virus
attenuated in cell culture (80).

The renal carcinoma of *Rana pipiens* was first described by
Lucké in 1934 (282*a*), and has since been known as Lucké's
adenocarcinoma. Four years later Lucké suggested the involvement of a virus in its aetiology (282*b*) although nearly another
20 years were to elapse before the resemblance of that agent to
viruses of the herpes group was recorded (151).

Application of the technique of cell culture has also allowed
an extension of the field of tumour-associated herpesviruses to
include a neoplastic disease of man, and so accentuated the
question of the possible oncogenicity of viruses of the herpes
group. The history of the disease concerned is brief and recent,
but the literature which has accumulated within the short span
of 15 years since it was first described by Burkitt (58) in 1962
bears witness to the intensity of the interest aroused, an interest
which has been stimulated by the complexity of the picture as
it evolved. When Burkitt first discovered the lymphoma named
after him in 1962, the climatic characteristics (high annual rain-
fall and high mean temperatures) of the tropical areas in which
the disease appeared to be endemic suggested to him a possible
viral aetiology and the involvement of an insect vector. Consequently an immediate and intense search was launched for
viruses associated with Burkitt's lymphoma, and the results were
not long in coming.

In studies published in the following two to three years, virus particles were found in biopsy material from Burkitt's lymphoma. They were initially described as being morphologically similar to herpesviruses but they could not be identified as any of the known members of this group (145). Named Epstein-Barr (EB) virus, the agent has subsequently been found to be identical with the pathogen responsible for the classic form of mononucleosis (209). Although its role in the pathogenesis of Burkitt's lymphoma is not yet clearly established, the peculiar affinity of EB virus for human lymphoid cells – to the exclusion of all other host cells, with the possible exception of some primate lymphoid cells – suggests that its viral genome may act specifically on lymphoid cells. On the other hand, it is also still possible that the association may be coincidental, and that the virus which in certain circumstances causes infectious mononucleosis may be carried by the neoplastic cells merely as a 'passenger' virus.

Thus, the extent of the involvement of herpesviruses in the pathogenesis of the tumours with which they are associated remains, for the time being, uncertain. There can be no doubt that the aetiology of Burkitt's lymphoma is a complex one, as are those of other conditions with tumour manifestations, and some of the virus-induced leukaemias. However, the very complexity of tumour virology makes it a focus of interest in more fields than one. As virology moves closer to immunology, the ramifications of both are felt in the study of oncogenesis; while at the same time developments in molecular biology throw light on other aspects of the same problems.

The observations made by Stewart and Eddy concerning the activities of polyoma virus in foreign hosts were noted by Dulbecco, who initially saw this small DNA virus as a tool in the study of genetic processes (cf. Chapter 9). Some of his young associates were already at work on Rous' sarcoma virus (439) when in 1958 Dulbecco launched a major study of polyoma virus, and later also of SV40, at the molecular level. The results obtained by his group at the California Institute of Technology, and later at the Salk Institute, profoundly influenced the understanding of the relationship between tumour viruses and their host cells and ultimately led to a revision of what had long been accepted as the 'central dogma' of molecular biology.[1] Throughout the 1960s Dulbecco and Marguerite Vogt and other associates who came and went studied the interactions between the DNA tumour viruses and their host cells. Their first discovery was the way in which cells became 'transformed' in infected cultures, showing morphological development and growth patterns which differed from those of uninfected cells. Such cell transformation (127) provided them with an *in vitro* system for the study of growth regulation changes in cells which were developing into malignant ones. Eventually Dulbecco and his group were able to establish the way in which the DNA of an invading virus becomes incorporated into the hereditary material of the transformed cell, and thus is enabled to code for proteins which control cell growth.

It was mentioned above that when Dulbecco first began work on polyoma virus in 1958–59 (130), some of his associates were already studying Rous' sarcoma virus. One of them was Howard Temin, whose interest in this particular virus was greatly stimulated by an observation he made after he moved to the University of Wisconsin in 1960. Temin found that infection of cells with Rous' sarcoma virus was blocked by inhibition of DNA synthesis, and of DNA-dependent RNA synthesis (437). Since the genome of Rous' sarcoma virus is known to be single-stranded RNA, this was an unexpected and well-nigh inexplicable finding. Temin concluded that a DNA intermediate must be involved in the process of infection, and postulated the existence of a DNA 'provirus' containing the genetic information of the RNA viral genome and responsible for the synthesis of the viral RNA progeny.

There was no enthusiastic welcome for Temin's provirus theory. It seemed too far-fetched, too much in opposition to all the tenets of molecular biology as they had become established since the launching by Watson and Crick of a proposed structure for DNA in 1953 (471). Nevertheless, by 1970 the sceptics had to admit defeat. Temin's theory was vindicated when he showed, in collaboration with Mizutani, that Rous' sarcoma virions carry

[1] R. Olby in *The path to the double helix* (335) has discussed the changing fortunes of the 'central dogma' and pointed out that Temin's results do not amount to an actual reversal of the central dogma. All the same, it shook accepted beliefs.

an enzyme capable of transcribing the single-stranded RNA of the virus into DNA (438). Their results were confirmed by simultaneous independent results obtained by David Baltimore, who had also worked at the Salk Institute in the 1960s. Baltimore demonstrated the presence of an RNA-dependent DNA polymerase in the virions of Rauscher murine leukaemia and of Rous' sarcoma (24). The two papers were published on successive pages of the same issue of *Nature* in June 1970.

This momentous discovery has led to new insight into the *modus operandi* of the tumour viruses. There is reason to suppose that the differences in this respect between DNA tumour viruses and their RNA counterparts may be trivial. They all effect the change into malignancy by becoming incorporated in the genetic material of the cells they are about to transform. Many other aspects of tumour virology remain in question, as do the differences and similarities between the mechanisms involved in viral oncogenesis and in chemically or physically induced carcinogenesis. Tumour virology is a field likely to go on expanding in the foreseeable future.

14. New light on long incubation periods

It has been pointed out in earlier chapters that rabies, because of the mysterious aspects of the disease, has come in for a great deal of attention from physician and layman alike through the ages. Even after Remlinger had filtered the pathogen in 1903 (p. 48) doubts about the aetiology returned when Negri described the inclusion bodies and considered them to be protozoal parasites (p. 49); and as late as the 1920s there were spirited discussions among pathologists as to the possible identity of similar parasites supposedly involved in the pathogenesis of rabies. The notion of spontaneous occurrence of the disease, prompted only by adverse environmental circumstances without the involvement of infection, also persisted until a very late date (3; 445). The doubts were finally and effectively removed when, by 1963, electron micrographs had been obtained of the virion, and its nucleic acid had been shown to be RNA (see 475).

However, even after these important facts had been established, one aspect of the disease remained as puzzling as ever: the long, sometimes extremely long, incubation period. For centuries, it had added to the mystery surrounding rabies infections; but where, in previous centuries, it had served to emphasize the unique nature of rabies compared with other infections, in the twentieth century evidence has accumulated to show that long incubation periods, as well as slowly developing manifestations with neurological involvement, are found in other diseases caused by filterable agents.

On 16 January 1943 the *Lancet* published a 'special article' entitled 'Homologous serum jaundice' (314). It was a memorandum prepared by medical officers of the British Ministry of Health, and it described outbreaks of hepatitis following (1) administration of measles convalescent serum to schoolchildren; (2) serum transfusion to hospital patients; and (3) administration of yellow fever vaccine. The report coincided with a rapid accumulation of data in this field in Britain and the United States, where from then on 'serum hepatitis' became the accepted term for parenterally transmitted, long incubation, viral hepatitis (296).

World War II proved to be a watershed in the history of serum hepatitis, not only in the question of terminology, but also by the overwhelming amount of evidence accumulated for the risk of transmission of infective hepatitis by different kinds of parenteral therapy with or without the use of human blood products. There was no watershed with regard to the prevention of the outbreaks, though. The report concluded: '. . . Although aetiological studies are proceeding both in England and America it is unlikely that the problem will be easily solved or that a radical method for preventing the phenomenon will readily be found.'

Infectious, epidemic forms of hepatic disease had by then been known for centuries. Early observations may be found in the works of Hippocrates (214). In the eighteenth century, outbreaks among troops were reported in Germany, and seasonal epidemic outbreaks were recorded by George Cleghorn when he was surgeon to the 22nd Regiment of foot soldiers stationed there from 1736 to 1749 (83).

In 1912, E. A. Cockayne wrote a well-documented study of 'catarrhal jaundice' (84). It included copious references and several case histories, one his own when he contracted hepatitis from a servant. Cockayne was able to distinguish between two or three distinct patterns of disease. One was what was then and still is known as Weil's disease, caused by a leptospire; but he also observed that there still appeared to be two distinct clinical courses in the remaining cases, and concluded: 'Thus I regard catarrhal jaundice as a specific disease due to an unknown organism endemic over a wide area and appearing not uncommonly in restricted and very rarely in widespread epidemics. If the view of its pathology which I have suggested is correct, the name 'infective hepatitis' would more accurately express the condition than 'catarrhal jaundice.'

Cockayne included among his references a paper by A. Lürman, describing an outbreak of jaundice following large-scale vaccination against smallpox of shipyard workers in Bremen (289). The vaccine used was based on 'humanized lymph', and the reported subsequent 191 cases of jaundice constituted the first recognizable record of serum hepatitis in the literature. The implications seem to have escaped Cockayne, who did not refer to it in his text, and in fact the paper was largely ignored for many years.[1]

In 1909, Paul Ehrlich, with Hata, introduced that outstanding product of his 'chemotherapy', the arsenical compound 606, better known as Salvarsan (see 136). It was the first synthetic drug to be widely used, and its impact on syphilis statistics was dramatic. The large proportion of secondary cases of jaundice was at first attributed solely to toxic effects of arsenic; but gradually, during the early 1920s, as hepatitis became increasingly common following Salvarsan treatment of luetic patients, it was suggested that an infectious agent might be responsible (430).

However, a true interpretation of the nature of serum hepatitis, and a suggestion that its aetiology might be different from that of infectious hepatitis, first came from Sweden. An outbreak in a diabetic clinic in 1926 led to the implication of infected needles and syringes, and in the same study proper attention was given, for the first time, to the long incubation period (155). The paper was published in German in a Swedish journal, and seems to have attracted less than its proper share of attention at the time.

In the late 1930s, in Britain, it was noted that outbreaks of hepatitis in some cases followed immunization against yellow fever (154). Then, in 1939, war broke out. The results of that war were disastrous to civilization at all levels. Nevertheless, there were areas where considerable insight was gained not just in the face of calamity, but by the very nature of that calamity. One such area was viral hepatitis, where suggestive evidence accumulated rapidly, both in Britain and in the United States,

---

[1] In 1918, Arnold Theiler, who had continued working on African horse sickness since the early filter experiments (p. 36), quite clearly encountered an outbreak of horse serum hepatitis in horses during trials with a vaccine against African horse sickness (441).

of outbreaks following (1) arsenical therapy of army syphilitics (450); (2) massive transfusions of blood and/or serum to casualties (if they survived long enough); and (3) the administration to troops on a very large scale of yellow fever vaccine, produced by mixing live yellow fever virus with human convalescent serum (396). These experiences also highlighted the particular risks associated with the use of pooled serum or plasma as compared with transfusion of whole blood from individual donors.

When the war came to a close, it was reasonably well established that there were two distinct kinds of viral hepatitis. They came to be known as hepatitis A, or infectious hepatitis, and hepatitis B, or serum hepatitis. Yet attempts to isolate their agents proved futile; and in spite of the considerable knowledge which had been gained concerning the spread of serum hepatitis, all efforts to control it remained unsuccessful. No real progress was made until the 1960s.

Towards the end of the previous decade, medical practice had received a bonus from what has been unkindly referred to as our plastic age. Metal (or glass and metal) syringes had been used to administer intravenous medication since the days of Harvey.[1] Durable and relatively expensive, they could be used indefinitely. Sterilization in between uses was not always adequate, and many outbreaks of serum hepatitis had been traced to infected syringes. Until, that is, the arrival of the disposable plastic syringe. It is not possible to give an exact date, not even to the year, for its introduction. It became available commercially in the late 1950s, and in Britain its progress is reflected in letters and short announcements in the *Lancet* and the *British medical journal*. By 1964–65 the disposable plastic syringe had truly arrived (281).

The use of plastic syringes has been an important help in the many-sided battle to control serum hepatitis; but in the history of this infection the year 1964 is chiefly distinguished by Baruch Blumberg's discovery of Australia antigen (47). After some initial confusion its presence was linked to serum hepatitis as opposed to infectious hepatitis (360); and although some aspects of this phenomenon have remained obscure, it has proved to

---

[1] For other purposes they had been in use even longer – a bronze syringe was described by Hero of Alexandria in the first century A.D.

Fig. 27. Demonstrating the technique of intravenous injection in the seventeenth century. From *Clysmatica nova*, Berlin: G. Schultz for D. Reichel, 1667. (Reproduced by courtesy of the Wellcome Trustees.)

172

be of immense value diagnostically and in the efforts to screen patients and staff at risk. More than 10 years after the discovery of Australia antigen, a successful outcome to the prolonged search for the agents of the viral hepatitides at last appears to be imminent (469); but the problem of developing an effective vaccine remains. The enormous complexity of the problem is well illustrated by a comparison of the cautious statement made recently (1972) by Maycock (312) that the discovery of Australia antigen 'began a series of advances which have culminated in giving transfusion services a means of exercising some degree of direct control over the transmission of hepatitis by blood and blood products' with the hopeful words written by Stokes nearly 20 years before (1953): '. . . we have means at least of controlling completely sharp outbreaks of epidemic hepatitis, and the complete control of the hepatitis virus (or viruses) B in plasma at least appears imminent. . .' (428).

With the observations of virus-like particles in electron micrographs of material from viral hepatitis A and B, some puzzling aspects of their aetiology are moving closer to a solution. In the meantime, other problems have come into focus. They include a group of diseases with an incubation period even longer than that of serum hepatitis and with manifestations developing so slowly that there is no clear line of demarcation between incubation period and onset of clinical disease. The group at present includes four conditions, i.e. scrapie of sheep, mink encephalopathy, kuru and Creutzfeldt–Jakob disease. There is good reason for their joint classification. Not only are they all characterized by involvement of the central nervous system as the respective diseases develop in their respective hosts; but their agents, which have not been isolated, have been shown to pass through 220-nm filters in transmission experiments. Yet they do not fit the conventional picture of a virus because their resistance to heat as well as to various kinds of radiation is much higher than that of any known virus. This would suggest that they differ from other known viruses in their content of nucleic acid.

Of the four diseases mentioned only one, scrapie, had been described before the present century. It has been recorded in Britain, France, Germany and Spain for more than 200 years, when it was shown in France to be transmissible by injection of filtered material from the brain and spinal cord of affected animals (97). After World War II experiments carried out in Britain confirmed and extended the French results, drawing attention to the unusual resistance of the agent, and to the surprisingly long incubation period (423). In 1954, Sigurdsson, in a classic paper, pointed out the close similarity of scrapie to the sheep disease called rida in Iceland, and introduced the term 'slow virus' infections which has since come into general use (414).

Shortly afterwards, on the other side of the globe, a medical officer in New Guinea observed what appeared to be the endemic occurrence, within a small, limited area of his district, of a degenerative and eventually fatal disease in the native population. When Sir Macfarlane Burnet arrived in the island on a visit, Zigas, the medical officer in question, consulted him about the disturbing outbreak. At that time, Burnet had a postdoctoral visitor from the United States working with him on immune mechanisms in his department at the Walter and Eliza Hall Institute in Melbourne. A bachelor with no home ties and commitments, but with a facility for languages and tremendous enthusiasm for a variety of subjects, Gajdusek was the ideal man for the study of a neurological disorder in a primitive tribe with cannibalistic tendencies in a remote tropical region. Combining anthropological and medical research, he spent 10 months living with the affected tribe, exploring their jungle existence and customs, making occasional forays into neighbouring areas to learn about the habits of other tribes and about the geographical distribution of the disease. By 1957, Gajdusek was back in the United States, and he and Zigas published the results of their investigations. In spite of the extensive work, the conclusions reached were vague and largely negative. There were probably 'strong genetic factors operating in pathogenesis' wrote Gajdusek and Zigas, while they had failed to find any effect of dietary factors, skin painting, smoke or other possible agents of organic or trace-metal character (169).

The problems posed by kuru continued to intrigue Gajdusek. They were also noted by others, and in 1959 the *Lancet* published a letter from William Hadlow (191) in which he drew attention to striking similarities between the pathology of kuru and that of scrapie. Hadlow was familiar with the successful transmission experiments performed with filtered suspensions of scrapie material, and suggested a similar approach for kuru using sub-human primates as experimental animals and observing them for extended periods. The implications were not lost on Gajdusek, and he very soon acted on Hadlow's suggestions. By 1963, the first chimpanzees were succumbing to kuru (168). Encouraged by their results, Gajdusek and his colleagues then repeated their chimpanzee transmission experiments with another slowly developing condition characterized by degenerative changes in the central nervous system. This had first been described in Germany by H. G. Creutzfeldt (94) and by A. Jakob (232) in 1920 and 1921, respectively, and had since been considered a heredo-familial pre-senile dementia with no suggestion that an infectious agent might be involved in its pathogenesis. In 1968, Gajdusek's team could report the successful transmission of Creutzfeldt–Jakob disease to chimpanzees. Shortly afterwards they were able to transmit both kuru and Creutzfeldt–Jakob disease to other, more easily available, primates.[1] The similarity seen in the clinical course of the diseases as well as in the cellular pathology of brain material from cases of kuru, Creutzfeldt–Jakob disease and scrapie has led to the assumption that the causative agents involved must be of a similar nature. Since scrapie[2] can be transmitted to mice, it lends itself more easily to experimental study.

The unusual characteristics of the scrapie agent have puzzled observers in much the same way as the filterable viruses did when they were first isolated at the turn of the century. From the beginning, a plant virus served as a useful analogue for its human and animal counterparts. Recent developments in plant pathology suggest that history may once again repeat itself, and that a plant analogue may ultimately assist in the exploration of the nature of the slow virus diseases. Since 1967, T. O. Diener and his group have been studying a number of plant pathogens

[1] The primates found to be susceptible include spider monkeys, squirrel monkeys and woolly monkeys (168); in the case of Creutzfeldt–Jakob disease it has also been transmitted to cats.
[2] There have been some indications that mink encephalopathy may in fact be identical with scrapie (468).

which have become known as viroids, and which exhibit many of the same peculiarities characterizing the scrapie agent (110; 111). Diener has suggested that their properties would accord with a composition of free nucleic acid; they are certainly very small indeed. As the dates indicate, this is not only an unfinished chapter, it is very much history in the making. It justifies inclusion in the present volume only as a pointer to one of several areas where at the moment virology is expanding fast. Another is the field of persistent virus infections, where the classic example is the perennial puzzle of herpes simplex infections (378); and perched uneasily between the persistent and the slow virus infections can be found an assortment of conditions including virus-induced tumours, sub-acute sclerosing panencephalitis and multiple sclerosis.[1] At the other end of the scale, there are virus infections not previously encountered which have taken the world by surprise in recent years: Marburg disease, Lassa fever and Ebola virus disease are the latest to strike on the African continent (415). It is a measure of the progress made in less than a century that electron micrographs of Ebola virus were obtained less than six months after the initial outbreak of the disease. On the other hand, looking at virus diseases throughout the whole spectrum from the slow virus diseases to the violently acute ones, it also serves to remind us that their history is far from complete.

[1] Very recently, there have been reports of a possible viral aetiology for a slowly developing nephropathy endemic in certain limited areas of the Balkans (16).

Complementary reading and general reference

Andrewes, Sir Christopher (1973). *In pursuit of the common cold.* London: William Heinemann Medical Books Ltd.

Andrewes, Sir Christopher and Pereira, H. G. (1972). *Viruses of vertebrates,* 3rd edn. London: Baillière, Tindall and Cassell.

Bawden, F. C. (1964). *Plant viruses and virus diseases,* 3rd edn. N.Y.: Ronald Press Co.

Bulloch, W. (1938). *The history of bacteriology.* London: Oxford Univ. Press.

Burnet, Sir Macfarlane (1968). *Changing patterns.* Melbourne and London: William Heinemann.

Burnet, Sir Macfarlane and White, D. O. (1972). *Natural history of infectious disease.* Cambridge: Cambridge Univ. Press.

Cairns, J., Stent, G. S. and Watson, J. D. (eds.) (1966). *Phage and the origins of molecular biology.* N.Y.: Cold Spring Harbor Lab. Quant. Biol.

Dixon, C. W. (1962). *Smallpox.* London: J. and A. Churchill Ltd.

Doerr, R. and Hallauer, C. (eds.) (1938). *Handbuch der Virusforschung,* vols. I and II. Wien: Verlag von Julius Springer.

Fenner, F. and Ratcliffe, F. N. (1965). *Myxomatosis.* London and N.Y.: Cambridge Univ. Press.

Gibbs, A. J. (ed.) (1973). *Viruses and invertebrates.* Amsterdam and London: North-Holland Publishing Co.

Hughes, S. S. (1977). *The virus: a history of the concept.* N.Y. and London: Heinemann Educational Books, Science History Publications.

Luria, S. E. (1953). *General virology,* 1st edn. N.Y.: John Wiley and Sons; London: Chapman and Hall Ltd.

Monod, Jaques and Borek, Ernest (eds.) (1971). *Of microbes and life*. N.Y. and London: Columbia Univ. Press.

Paul, J. R. (1971). *A history of poliomyelitis*. New Haven and London: Yale Univ. Press.

Rivers, Thomas M. (ed.) (1928). *Filterable viruses*. Baltimore: Williams and Wilkins Co.

Stent, Gunther S. (1963). *Molecular biology of bacterial viruses*. San Francisco and London: W. H. Freeman and Co.

Biographical notes

References at the end of entries refer to obituaries or articles in which additional biographical information may be found.

Abbé, Ernst   1840–1905
   Educated at the Universities of Jena and Göttingen, Abbé taught physics, mathematics and astronomy at Jena, from 1870 as professor, and later was also director of the astronomical and meteorological observatories. In Jena he became associated with Carl Zeiss of the optical instruments firm, from 1866 as research director there, and after the death of Zeiss in 1888 he took over the firm's direction. In 1891 Abbé established and endowed the Carl Zeiss Foundation for research in science and social improvement; five years later he introduced worker participation, reorganizing the works into a cooperative in which management, workers and the university shared in the profits. An acknowledged authority on theoretical aspects of the microscope, Abbé introduced such improvements in design as the use of condensers to give strong, even illumination, and homogeneous immersion lenses.

Andrewes, Sir Christopher [Howard] (C. H. Andrewes, knighted 1961)   1896–
FRS 1939
   Born in London in 1896, Andrewes studied medicine at St Bartholomew's Hospital where his father (Sir Frederick William Andrewes) was Professor of Pathology. Influenced by the enthusiasm of Mervyn Gordon, Andrewes developed an interest in work on vaccinia and mumps. After a period at the Rockefeller Institute, where he worked on virus III and met Rivers and Peyton Rous, he returned to London and eventually settled for laboratory research as opposed to clinical medicine, and pioneered work on the transmission of influenza together with Wilson Smith and Laidlaw. After World War II he established the Common Cold Research Centre at Salisbury, and worked on classification and nomenclature; on the brink of retirement he wrote the classic reference work *Viruses of vertebrates* (later editions in collaboration with H. G. Pereira).
   (*A. Rev. Microbiol.*, 1973, **27**: 1–11.)

Avery, Oswald Theodore   1877–1955   For.M.R.S. 1944
   A graduate of Columbia University, New York, Avery became the inspired leader at the Rockefeller Institute of a group studying capsular polysac-

charides of pneumococci. In 1923 he interested Heidelberger and others in the chemical differences between pneumococci of different antigenic structure, opening up a whole new field of serology. An exceptionally generous and unselfish scholar, his very modesty plus adverse facts of war and publishing all contributed to the delay in recognition of his discovery of DNA as the carrier of genetic information. Dubos has suggested that the Nobel Committee should recognize, as the Académie Française did for Molière, that '*Rien ne manquait à sa gloire; Il manquait à la nôtre*' (*Biogr. Mem. Fellows R. Soc.*, 1956, **2**: 35–48). *Nobel: the man and his prizes*, edited by the Nobel Foundation, contains an apologia: 'Thus, Avery's discovery in 1944 of DNA as carrier of heredity represents one of the most important achievements in genetics, and it is to be regretted that he did not receive the Nobel Prize. By the time dissident voices were silenced, he had passed away'.
(*Biogr. Mem. natn. Acad. Sci.*, 1958, **32**: 31–49.)

Bang, Oluf   1881–1937
Born in Copenhagen, the son of Professor Bernhard Bang who pioneered research on cattle tuberculosis and *Brucella abortus*, Bang graduated in veterinary science in 1905, and became professor of veterinary medicine in 1914. In 1908 he published with Ellerman the classic study of chicken leukaemia.

Barnard, Joseph Edwin   1870–1949   FRS 1924
The son of a St James' Street hatter, Barnard was a late example of the Victorian amateur microscopist who rose through the ranks of the Royal Microscopical Society to be elected a Fellow of the Royal Society for his outstanding contributions in the field of ultraviolet microscopy, which culminated in his studies on a number of filterable viruses.
(*Obit. Not. Fell. R. Soc. Lond.*, 1950–51, **7**: 3–8.)

Bawden, Sir Frederick [Charles] (F. C. Bawden, kt. 1967)   1908–72   FRS 1949
Educated at Cambridge, in 1930 he began plant virus research at Salaman's Potato Virus Research Station at Cambridge. He carried out epoch-making work on tobacco mosaic and other plant viruses with N. W. Pirie from the mid-1930s. In 1936, Bawden moved to Rothamsted as 'virus physiologist'. He was director of Rothamsted Experimental Station from 1958.
(*Biogr. Mem. Fellows R. Soc.*, 1973, **19**: 19–63.)

Bechhold, Heinrich   1866–?
Educated at Universities of Freiburg, Strasbourg, Berlin and Heidelberg, he worked on colloids from the early days of colloid chemistry, wrote prolifically on the subject, and devised ultrafiltration methods as a tool for their study, especially in relation to medicine (agglutination of bacteria, disinfection, colloid therapeutics, sizes of viruses and bacteriophage). He published the first edition of *Die Kolloide in Biologie und Medizin* 1912 (5th edn, 1929).

Bedson, Sir Samuel [Phillips] (S. P. Bedson, knighted 1956)   1886–1969   FRS 1935
Born at Newcastle-upon-Tyne, he was the son of a Professor of Chemistry, who insisted that he took a B.Sc. in chemistry before going on to graduate

in medicine at Durham University in 1912. After a period at the Institut Pasteur in Paris, and subsequently at the Lister Institute, London, Bedson's career was interrupted by World War I service. When enlisting, Bedson was told 'No vacancies in RAMC for doctors trained in research', but after gallant fighting as combatant officer receiving severe chest wound he was finally transferred as pathologist to RAMC in France. As research fellow at the London Hospital in 1929 he became interested in cases treated there of psittacosis transmitted from imported parrots and parakeets, and wrote a classic paper in 1936 on this class of pathogens (*Bedsonia*) with which his name has become permanently associated.

(*Biogr. Mem. Fellows R. Soc.*, 1970, **16**: 15–35.)

Beijerinck, Martinus Willem   1851–1931   For.M.R.S. 1926
Dutch microbiologist. Born at Amsterdam, Beijerinck was educated at the Delft Polytechnic School and the University of Leiden, where his room mate (and later life-long friend) Jan van't Hoff of stereochemistry fame influenced his thinking. After graduation Beijerinck went to teach botany at Wageningen, where Adolf Mayer introduced him to work on tobacco mosaic disease. Beijerinck made a number of fundamental contributions to the field of microbiology, notably the isolation of the nitrogen-fixing bacterium *Rhizobium leguminosarum* and the development of enrichment culture methods, but to the virologist the main focus of interest is his work on the pathogenesis of tobacco mosaic disease.

(*Proc. R. Soc. Lond.* B, 1932, **109**: i–iii.)

Bernal, John Desmond   1901–71   FRS 1937
Born in Ireland of an American college graduate, protestant mother and a Catholic gentleman farmer father, educated at an English Public School and Cambridge, the ruling passions of Bernal's life were to be Roman catholicism, Irish nationalism, mathematics and the physical sciences. At Cambridge he was attracted to X-ray crystallography and the circle around William Bragg, in particular Kathleen Lonsdale and W. T. Astbury, and later Dorothy Hodgkin, Fankuchen and Perutz. In addition to much early and important work on X-ray crystallography of viruses, Bernal was personal adviser to Lord Mountbatten during the war, and published in 1939 a volume on *The social functions of science.*

(*Nature*, 1972, **235**: 235–6.)

Blumberg, Baruch Samuel   1925–
Born in 1925, Blumberg was educated in New York, graduating M.D. from Columbia University in 1951 and adding a Ph.D. six years later after spending 1955–57 at Oxford. At the Institute for Cancer Research in Philadelphia, he discovered the Australia antigen (Hepatitis B antigen) in the serum of an Australian aborigine in 1963. He was awarded the Nobel Prize jointly with D. C. Gajdusek in 1976.

von Bollinger, Otto   1843–1909
German pathologist, educated at Munich, Vienna and Berlin; in 1874 he became professor at a Munich Veterinary School, and in 1880 Professor

of Pathology and Pathological Anatomy at the University of Munich. Von Bollinger's particular interests were actinomycosis, anthrax and tuberculosis, and in 1873 he published observations on the inclusion bodies associated with fowl pox.

Bordet, Jules Jean Baptiste Vincent   1870–1961   For.M.R.S. 1916
    The son of a schoolmaster, Bordet was educated at the Free University of Brussels. In his final year of medical studies he published his first paper, on bacterial defense against phagocytosis. From 1894–1901 he was at the Institut Pasteur in Paris except for a period spent with A. Theiler in the Transvaal to study rinderpest, and taking part in Theiler's investigations into the problems posed by African horse sickness. In 1900, Bordet was appointed director of the newly established Institut Antirabique et Bactériologique de Brabant, later to become the Institut Pasteur de Bruxelles. In 1919, Bordet was awarded the first post-war Nobel Prize for Physiology or Medicine for 'his discoveries relating to immunity'. From 1920 onwards, Bordet became involved in the study of bacteriophage, on a number of occasions sharply opposing d'Herelle.
    (*Biogr. Mem. Fellows R. Soc.*, 1962, **8**: 19–25.)

Borrel, Amédée   1867–1936
    Graduating in medicine at the University of Montpellier, Borrel immediately entered the laboratory of Metchnikoff in 1892. After many years spent at the Institut Pasteur in Paris, Borrel eventually accepted the directorship of the Institute of Hygiene at Strasbourg. His work reflects all the major problems concerning the Institut Pasteur during his lifetime, such as phagocytosis, filterable viruses, cytology and parasitology of tumours, bacteriophage and tissue culture; but within all these areas he focused always on the aspects affecting his consuming interest – the problem of cancer.
    (*Annls Inst. Pasteur, Paris*, 1936, **57**: 337–42.)

Boycott, Arthur Edwin   1877–1938   FRS 1914
    Interested in natural science since boyhood, Boycott went up to Oxford as a Senior Classical Scholar but soon changed to physiology in spite of opposition from his tutors, and was introduced to research by Burdon-Sanderson. After St Thomas' Hospital Medical School and a D.M. at Oxford in 1904, he was appointed to the Chair of Pathology at University College, London, in 1914. His wide range of interests was reflected in numerous papers on physiological, pathological and zoological subjects, and he was for many years editor of the *Journal of Pathology and Bacteriology*. In his later years he turned his attention increasingly to theoretical matters, as evidenced by his speculations on the nature of filterable viruses.
    (*Obit. Not. Fell. R. Soc. Lond.*, 1936–38, **2**: 561–71.)

Buist, John Brown   1846–1915   A farmer's son from Fife, Buist received his medical education at the University of Edinburgh and the Royal College of surgeons in the 1860s. Although he lectured for a time on general pathology he devoted most of his working life to vaccination, becoming a Local Government Board teacher of vaccination. Dedicated to his work, Buist died while disregarding his own cardiac condition in order to fulfill his duties as vaccinator. His best-known work is his writings on the 'micro-organisms' of vaccinia and variola published in 1886 and 1887, which include his observations of the stained elementary bodies of vaccinia.

(*Br. med. J.*, 1915, i: 274.)

Bulloch, William E.   1868–1941   FRS 1913
Born and educated at Aberdeen, Bulloch became bacteriologist to the London Hospital and Lecturer on Bacteriology and Pathological Chemistry in its Medical School in 1897. Remaining there until his retirement in 1934, he received the title of Goldsmiths' Company Professor of Bacteriology in 1919. A revered teacher and prolific writer, Bulloch published in 1938 the classic volume *The history of bacteriology.*

(*Obit. Not. Fell. R. Soc. Lond.*, 1939–41, **3**: 819–52.)

Burdon-Sanderson, Sir John Scott   1828–1905 (baronet 1899)   FRS 1867
Born in Northumberland, Burdon-Sanderson received his early education at home, where his father wished him to follow his own profession of the law; but conversation with a doctor who was treating his brother during a serious illness made him decide to study medicine instead. Having graduated from Edinburgh University he became Medical Officer of Health at Paddington and wrote on contagion in Privy Council Reports. He was appointed Professor of Physiology at University College, London, in 1874 and at Oxford in 1883, and completed his life's work as Regius Professor of Medicine at Oxford from 1895 to 1905.

(*Br. med. J.*, 1905, ii: 1481–95.)

Burnet, Sir Macfarlane (F. M. Burnet, knighted 1951)   FRS 1942
Born in 1899, Burnet was educated at Melbourne University which he was later to serve as Professor of Experimental Medicine and Director of the Walter and Eliza Hall Institute for Medical Research. A visitor at the National Institute for Medical Research in London in the early 1930s during the heyday of influenza research, Burnet has a number of discoveries to his credit in a variety of fields, such as influenza and other myxoviruses, haemagglutination, bacteriophage, etc. In the early 1950s he turned his attention, and with it the direction of the Walter and Eliza Hall Institute, more exclusively towards the rapidly developing discipline of immunology. He developed the theory of clonal selection of antibodies and was awarded the Nobel Prize, jointly with Sir Peter Medawar, for 'discovery of acquired immunological tolerance' in 1960.

Carrel, Alexis   1873–1944
Born near Lyon, Carrel took his medical degree at the University of Lyon in 1900. After staying there for two years he left France for the United States where he worked at the University of Chicago for another two years before joining the staff of the Rockefeller Institute, and it was here that he carried out much of the work for which he has become famous. His development of the artificial heart in collaboration with Charles Lindbergh received much publicity, while in the history of virology his tissue culture work is of more

interest. Carrel was awarded the Nobel Prize in 1912 for work on 'vascular suture and the transplantation of blood vessels and organs'.

Carroll, James 1854–1907
  Born in England, Carrol emigrated to Canada at the age of 15, living as a backwoodsman until he enlisted in the United States Army in 1881 and, still a soldier, began the study of medicine in New York in 1886–87. He eventually graduated from the University of Maryland and worked at Johns Hopkins Hospital in pathology and bacteriology until beginning his association with Walter Reed and yellow fever research through the Army Medical School in Washington and the Yellow Fever Commission.

Centanni, Eugenio 1863–1942
  Italian pathologist who worked in northern Italy, first at Ferrara, then as Professor of Pathology at Modena, and from 1929 in Bologna. Having discovered the filterable virus of fowl plague in 1901, he was an early adherent, with his compatriot Sanfelice, of the idea of the chemical nature of the virus.
  (*Pathologica*, 1942, **34**: 345–7.)

Chamberland, Charles 1851–1908
  Having graduated from the rigorous Ecole Normale as a physiological chemist, Chamberland joined Pasteur's laboratory at the height of the spontaneous generation controversy, and in his thesis established the rules for sterilization of culture media. A robust Victorian full of joie de vivre, Chamberland was an invaluable collaborator for Pasteur and Roux in rabies and other research, and in the development of the Pasteur–Chamberland filters.
  (*Annls Inst. Pasteur, Paris* 1908, **22**: 369–74.)

Chauveau, Jean-Baptiste Auguste 1827–1917
  French pathologist and physiologist, educated at the Veterinary School of Alfort, and, after completion of his studies there, working at the Veterinary School at Lyon, where he pursued two very different lines of research: the development of a method of intra-cardiac exploration in the horse, and the study of the nature of the infective principle in vaccinia lymph. From 1875 he was Director of the Veterinary School at Lyon; his research on vaccinia earned him a doctorate in medicine from the Medical Faculty of Paris in 1877, and later in the same year he was called to the Chair of Experimental and Comparative Medicine in the Faculty of Medicine at Lyon. In 1886 he moved to Paris as Inspector General of Veterinary Schools and Professor of Comparative Pathology at the Museum of Natural History. In 1868, Chauveau published a series of classic papers on the nature and infectivity of the virus of vaccinia, in which he introduced the term 'elementary bodies'.

Copeman, Sydney Arthur Monckton 1862–1947 FRS 1903
  English pathologist, educated at Cambridge and St Thomas' Hospital Medical School, Copeman is known for his studies of variola and vaccinia, and in particular the invention of glycerinated lymph. The addition of glycerol to vaccinia lymph as a means of preserving it had been practised before by others,

but Copeman established the principle of selective bactericidal action by the added glycerol, and tirelessly promoted this improvement. He published *Vaccination; its natural history and pathology* in 1898.

(*Obit. Not. Fell. R. Soc. Lond.*, 1948–49, **6**: 37–50.)

Delbrück, Max   1906–   For.M.R.S. 1967

Born in Berlin and brought up in an academic family – Delbrück's father was a Professor of History, his mother a grand-daughter of Justus von Liebig – his boyhood interest was astronomy, deflected towards physical chemistry by his older friend K. F. Bonhoeffer, and later, during his student years, turning from astrophysics to theoretical physics in the heyday of development in quantum mechanics. During postdoctoral years abroad in the early 1930s, Delbrück was influenced by Wolfgang Pauli and Niels Bohr, and after a period under Lise Meitner he joined the group around Timoféeff-Ressovsky in the Berlin Kaiser Wilhelm Institutes, and began his long years of study of genetic mechanisms, emigrating to the United States on the eve of World War II. At the California Institute of Technology since 1947, Delbrück was one of the founders and leading lights of the post-war school of molecular biology. Since the early 1950s his interests have focused on sensory physiology and the study of *Phycomyces* as a model system for stimulus transduction. He received the Nobel Prize, jointly with Luria and Hershey, in 1969 for 'discoveries concerning the replication mechanism and the genetic structure of viruses'.

D'Herelle, Felix Hubert   1873–1949

Born in Montreal, Canada, d'Herelle lost his father early and was taken by his mother to Paris at the age of six. Educated in Paris and later at Medical School in Leiden, he became director of bacteriological laboratories in Guatemala and Peru in the first decade of the twentieth century. In 1908, d'Herelle joined the Institut Pasteur in Paris. He described bacteriophagy in 1917. In 1922 he left Paris for a Chair in Leiden, and from 1923 to 1927 directed the International Sanitary Commission in Egypt. From 1928 he occupied the Chair of Protobiology at Yale University. From 1917 to his death in 1949 d'Herelle tirelessly defended his view of bacteriophage as an organism parasitizing bacteria.

(*Annls Inst. Pasteur, Paris*, 1949, **76**: 457–60.)

Doerr, Robert   1871–1952

Born in Hungary, Doerr graduated in medicine at Vienna in 1897 and became attached to the Institute of Pathological Anatomy there. At the end of World War I he accepted the Chair of Hygiene and Bacteriology at the University of Basle. Doerr studied a number of filterable viruses since the early years of the twentieth century in addition to his investigations of immunological problems and allergy mechanisms. He co-edited and contributed to *Handbuch der Virusforschung*, 1938.

Duclaux, Pierre Emile   1840–1904

Born in the Auvergne, Duclaux was brought up in a strict, bourgeois French fashion, well taught by a father who was said to be a 'second James Mill or Etienne Pascal'. Well versed in French classical literature he eventually chose science and at Paris was taught by Pasteur. At the age of 26 he became Professor of Chemistry first at Tours, then Clermont-Ferrand; subsequently he spent five years as Professor of Physics at Lyon, and from 1878 held the Chair of Meteorology in the Institute of Agronomy in Paris. When the Institut Pasteur was founded in 1888 Duclaux joined the staff and eventually became Pasteur's first successor as director. He wrote *Ferments et maladies*, 1882 (dedicated to his wife who had died of puerperal fever), *Traité de microbiologie*, 4 vols, 1898–1900, and *Pasteur: histoire d'un esprit*, 1896.

Duggar, Benjamin Minge   1872–1956
Born in Alabama, Duggar graduated from Harvard in 1895 and travelled widely in Europe before working for 15 years at Washington University and the Missouri Botanical Garden. In 1927 he was appointed to the Chair of Plant Physiology and Economic Botany at the University of Wisconsin where he remained until he retired in 1943. As a very active Emeritus, Duggar became consultant in mycological research to the firm of Lederle, working on anti-malarial drugs, and discovered aureomycin in 1948.
   (*Biogr. Mem. natn. Acad. Sci.*, 1958, **32**: 113–31.)

Dulbecco, Renato   1914–   For.M.R.S. 1974
Born in 1914, Dulbecco graduated from the Medical School of the University of Turin and worked for a while in the pathological laboratories there at the same time as his slightly older contemporary, S. E. Luria, until the advent of World War II. Wounded on the Russian front in 1942, he returned to Italy and joined the Resistance. The end of the war saw him back in the Turin Pathology Laboratory until, at the instigation of Luria, he came to the United States where he soon joined Max Delbrück at CalTech. In 1962 Dulbecco moved to the Salk Institute, and in 1972 to London. Known for his work on bacteriophage, tumour viruses and development of the animal virus plaque-counting technique, he was awarded the Nobel Prize, jointly with D. Baltimore and H. Temin, in 1975 for 'discovery of interactions between tumour viruses and the genetic material of the host cell'.

Elford, William Joseph   1900–52   FRS 1950
Having graduated in chemistry from the University of Bristol in 1923, Elford in 1925 came to work at the National Institute for Medical Research in London under J. E. Barnard, who at that time was involved in the development of ultraviolet microscopy, and who was a friend of Bechhold. Elford soon became absorbed in work on ultrafiltration and made major contributions to the development of the graded collodion membranes which at that time provided the guide to the sizes of viruses.
   (*Obit. Not. Fell. R. Soc. Lond.*, 1952–53, **8**: 149–58.)

Ellerman, Vilhelm   1871–1924
Ellerman graduated in medicine from the University of Copenhagen in 1896, and in subsequent years studied at Heidelberg, Paris, Nancy and Berlin until taking up posts in Copenhagen hospitals. Gravitating towards forensic medicine, he was appointed to the Chair of Forensic Medicine at the University

of Copenhagen in 1914. He published a number of papers on transmissible leucosis of the fowl, the first with O. Bang in 1908, in which they described a leukaemia of chickens caused by a filterable virus.

Enders, John Franklin 1897–
    Born in 1897, Enders' studies at Yale University were interrupted by service in World War I. After graduating A.B. from Yale in 1920, he embarked on the study of English literature at Harvard, but soon changed to medicine. Impressed by Zinsser's teaching he then decided to pursue a Ph.D. course in microbiology and give up medicine (to the dismay of his family). In Zinsser's department he began studying viruses and their growth in eggs. After World War II, in the laboratories of the Boston Children's Hospital, Enders, with F. C. Robbins and T. H. Weller, was able to cultivate poliovirus in non-nervous tissue through improvements in tissue culture techniques. They also made important discoveries concerning quantitative determination of the virus by means of its cytopathic effects. Enders, Robbins and Weller were awarded the 1954 Nobel Prize for 'discovery of the ability of poliomyelitis viruses to grow in cultures of various types of tissue'.

Errera, Léo 1858–1905
    Belgian naturalist with a cosmopolitan background and catholic interests. His father was a Venetian banker who left his native city to avoid serving under the Austrian occupational regime and settled in Brussels where he married the daughter of a local banker. Botany was Errera's first and lasting love, but his intellectual capacities and mathematical ability gave him a wide range, and he wrote a number of books and articles dealing with theoretical aspects of biological subjects, as well as with more purely philosophical themes.
    (*Ber. dt. bot. Ges.*, 1905, **23**: 43–55.)

Fankuchen, Isadore 1904–64
    American biochemist who in 1934 came to Manchester to Bragg's group and later worked with Bernal at Cambridge studying the structures of plant viruses by means of X-ray diffraction and crystallographic techniques before returning to a Chair at the Brooklyn Polytechnic Institute in New York.
    (*Nature*, 1964, **203**: 916–17.)

Flexner, Simon 1863–1946 For.M.R.S. 1919
    American virologist. The son of a recent immigrant to the United States, Flexner began his studies in difficult financial circumstances and qualified in pharmacy before being able to turn to medicine. From a Chair of Pathological Anatomy at the Johns Hopkins University's Medical School, Flexner moved in 1903 to the newly established (1901) Rockefeller Institute of Medical Research as its Director. In this position he was invaluable, not only as administrator but as friend and mentor to all younger staff members. His rigidly held beliefs concerning aspects of the problem of poliomyelitis unfortunately arrested rather than furthered progress in this field during his directorship at the Rockefeller, but as editor of the *Journal of Experimental Medicine* Flexner also made a notable contribution.

Biographical notes

Fracastoro, Girolamo   c. 1478–1553
   Italian poet and physician. Educated at the University of Padua, Fracastoro taught there as Professor of Philosophy from 1502 (a colleague was Copernicus). Fracastoro's ideas on epidemic diseases and their spread were far in advance of their time. They were recorded in *De contagione et contagionis morbis et eorum curatione* (1546), where he declares his belief that epidemic diseases are caused by minute, multiplying bodies which are transmitted (1) by direct contact, (2) via soiled clothing and linen or (3) through the air. He is perhaps best known for his poem *Syphilis sive morbus Gallicus* (1530).

Fraenkel-Conrat, Heinz L.   1910–
   Born in Breslau in 1910, Fraenkel-Conrat was educated in Vienna, Munich, Geneva and Breslau and in 1936 in Edinburgh he added a Ph.D. in biochemistry to his medical degree, then left for the United States. In 1952 he joined the Virus Laboratory established for Stanley on the Berkeley campus of the University of California, where he is now Professor of Virology. Here he demonstrated in 1956, independently of and simultaneously with Gierer and Schramm, that the infectivity of the tobacco mosaic virus particle resides in its RNA.

Franklin, Rosalind E.   1920–58
   Having graduated in physical chemistry at Cambridge, Rosalind Franklin studied X-ray diffraction in Paris in the late 1940s, then worked on problems in biophysics at King's College, London. From 1953 she directed research on virus structure at Birkbeck College, London. She played a major role in a number of structural studies of DNA and of tobacco mosaic virus in the 1950s. (*Nature*, 1958, **182**: 154.)

Frosch, Paul   1860–1928
   Educated in Leipzig, Würzburg and Berlin, Frosch became assistant to Robert Koch and followed him on a number of expeditions abroad. He was made Director of the Research Department at the Institut für Infektionskrankheiten, and from 1897 was Professor of Bacteriology at the Veterinary College in Berlin. With Friedrich Loeffler, Frosch headed the commission set up to investigate foot-and-mouth disease in Germany, the work which resulted in its pathogen being recognized as the first filterable animal virus on record.

Fuller, Thomas   1654–1734
   English country physician, educated at Queen's College, Cambridge, where his studies of Descartes and Willis left a lasting impression. Graduated M.B. 1676 and M.D. 1681, after which he practised at Sevenoaks, Kent, for the rest of his life. By his own account, Fuller felt isolated in his country practice, but exercised his active mind by collecting, and publishing, several editions of pharmacopoeias of the 'best and neatest medicines', and *Exanthematologia*, a treatise on eruptive fevers, in 1730. In addition to this impressive output Fuller published, two years before his death, *Gnomologia: adagies, proverbs, wise sentences and witty sayings, ancient and modern, foreign and British*.
   (*Bull. Soc. med. Hist., Chicago*, 1917–22, **2**: 321–33.)

Biographical notes

Gajdusek, D. Carleton 1923–

Born in New York in 1923, Gajdusek graduated from the Harvard Medical School in 1946. After some time in clinical work in the United States he spent a year in the Department for Viral and Rickettsial Diseases at the Institut Pasteur in Iran, and in 1955 went to Australia, where the problem of kuru among the natives in the eastern Highlands of New Guinea led him to his pioneering research on this and other slow virus diseases. For this work he was awarded the Nobel Prize (jointly with B. Blumberg) in 1976.

Galtier, (Pierre) Victor 1846–1908

Of French peasant stock, Galtier entered the Veterinary School at Lyon at the age of 22, and subsequently spent all of his active working life there, from 1878 as Professor of Pathology. Concerned in particular with infectious diseases of domestic animals, Galtier made contributions to the study of glanders, tuberculosis, anthrax and above all rabies, for which he firmly established the use of the rabbit as experimental animal.

Gamgee, John 1831–94

The second son of a veterinary surgeon, Gamgee was born in Italy and lived for a time with his family in Florence. While his elder brother Joseph was educated to follow in their father's footsteps in veterinary practice, John Gamgee was destined for medicine; but the elder brother turned to human surgery after two years' veterinary practice, whereas John qualified at the Royal Veterinary College in 1852 without ever having completed his medical course. In 1857, Gamgee established the New Edinburgh Veterinary College which he moved to London in 1865, the year of the catastrophic outbreak of cattle plague in England. Gamgee had again and again warned the authorities about the approaching danger from the continent and suggested ways of prevention, to no avail. Embittered by the official attitude and the failure of his school through lack of support, Gamgee left for Texas where he investigated Texas fever for an American Government Commission, and also studied yellow fever. He returned to Britain in 1869, but forsook veterinary medicine for thermodynamics, and worked on improving methods for refrigerated transport of meat. He published *The cattle plague with official reports of the International Veterinary Congresses held in Hamburg, 1863, and in Vienna, 1865.*
(*Med. Hist.*, 1962, **6**: 45–58.)

Gatti, Angelo 1724–98

Italian physician, born in Tuscany, who from 1755 to 1762 was Professor of Medicine at Pisa. In 1761 he moved to Paris, where he popularized inoculation and wrote two volumes on the practice of inoculation and the possibilities, remote as they seemed at the time, of attenuating the active principle of variola for this purpose.

Gey, George Otto 1899–1970

Born in Pittsburgh, Gey graduated in natural science from the local university and for a while taught zoology before gaining an M.D. from the Medical School of Johns Hopkins University in 1933, where he eventually became director of the Tissue Culture Laboratory. Gey made many innovations in this field in which he worked for nearly 50 years, among them the 'roller tube' technique, 'flying cover slips' and simultaneous propagation of cells and viruses over long periods of time. He also introduced the HeLa continuous cell line derived from a cervical cancer.

Goodpasture, Ernest William   1886–1960
  Born on a farm in Tenessee, Goodpasture graduated M.D. from the Johns Hopkins University Medical School in 1912. In the early 1920s he became interested in work on herpesviruses, and as Professor of Pathology at Vanderbilt University he led a group working to establish Lipschütz's 'elementary bodies' as the specific pathogen of herpes. Subsequently they found fowl pox to provide a more promising model system and were able to identify the Borrel body as the pathogen. With Alice Woodruff, Goodpasture then attempted to cultivate the virus on the chorio-allantoic membrane of the developing chick embryo. The success of this experiment established egg culture as a most important culture method not just for fowl pox virus but for a considerable number of other viruses.
  (*Biogr. Mem. natn. Acad. Sci.*, 1965, **38**: 111–44.)

Gregg, Sir Norman [McAlister] (N. M. Gregg, knighted 1953)   1892–1966
  Graduating from the University of Sydney in 1915, Gregg went straight to France with the Australian forces in World War I. After the war he spent several years in England, returning to Sydney to build up a large ophthalmic practice. In October 1941 Gregg presented the epoch-making paper on 'Congenital cataract following German measles in the mother' which was to change attitudes throughout the world to what had hitherto been regarded as a mild, unimportant childhood disease. From 1925 until his death, Gregg was associated with the Royal Alexandra Hospital for Children in Sydney.
  (*Aust. paediat. J.*, 1966, **2**: 242–3.)

Guarnieri, Guiseppe   1856–1918
  Born in Pisa and educated in Rome, Guarnieri returned to Pisa as Professor of Pathology. He described the inclusion bodies in vaccinia and variola to be known by his name in 1892, comparing them to the bodies described in fowl pox by his colleague at Pisa, Rivolta, and by von Bollinger. He wrongly identified the bodies as protozoal parasites which he called *Citoryctes vaccinae* and *Citoryctes variolae*.

Gye, William Ewart (Before 1919, W. E. Bullock) 1884–1952   FRS 1938
  Originally a chemist, Gye came late to medicine, qualifying at the University of Edinburgh in 1912. In 1913 he joined the staff of the Imperial Cancer Research Fund in London; World War I intervened, and after service with the RAMC in France and Italy he returned to join the National Institute for Medical Research. In 1919 he also changed his surname from his father's name of Bullock to Gye, his first wife's maiden name, to 'avoid tiresome confusion' with William E. Bulloch. He wrote, with W. J. Purdy, *The cause of cancer*, 1931.
  (*Obit. Not. Fell. R. Soc., Lond.*, 1952–53, **8**: 419–30.)

Biographical notes

Hallauer, Curt 1900–

Swiss bacteriologist and virologist who was born in 1900 and educated at the Universities of Basle, Munich, Vienna and Berlin. From 1936 he was Professor at the University of Berne and Director of its Bacteriological Institute. He made early contributions to the development of tissue culture methods and suggested the possibility of virus titration based on cytopathic effects in 1931. He was co-editor and contributor to *Handbuch der Virusforschung* (1938).

(*Arch. ges. Virusforsch.*, 1970, **31**: i–iii.)

Hare, Ronald 1899–

Born in 1899 and graduating from St Mary's Hospital Medical School in 1924, Hare went straight into research in the Inoculation Department of St Mary's without serving an internship. He has engagingly described this unusual start to his career, and the dearth of encouragement for those who wished to involve themselves in bacteriological research in the period between the wars, in the largely autobiographical *The birth of Penicillin* (1970). In 1941, Hare, with McClelland, discovered the phenomenon of haemagglutination in influenza viruses, independently of and simultaneously with G. K. Hirst.

Harrison, Ross Granville 1870–1959 For.M.R.S. 1940

American anatomist and physician, educated at Johns Hopkins University where he later taught anatomy until becoming Bronson Professor of Comparative Anatomy at Yale in 1907. He had a period of further study in Germany before the turn of the century. Between 1907 and 1912 Harrison developed his method of cultivating nerve cells in extraneous media which enabled him to observe the growth of living nerve fibres and which formed the basis for the development of tissue culture techniques.

(*Biogr. Mem. Fellows R. Soc.*, 1961, **7**: 111–26.)

Hektoen, Ludvig 1863–1951

Born in Westby, Wisconsin, Hektoen's early years were 'those of a farm boy in a Norwegian community in which English was spoken only at school'. Friendship with a young Norwegian physician turned his thoughts to the study of medicine, and he graduated in Chicago in 1888. In 1898 he became Professor of Pathology at Rush Medical College, and in 1901 he was appointed to the same position in Chicago. From the early years of the century Hektoen was interested in the presence of measles virus in the blood of patients, and the possibility of developing a vaccine.

(*Biogr. Mem. natn. Acad. Sci.*, 1954, **28**: 163–97.)

Hershey, Alfred Day 1908–

American molecular geneticist, born in 1908, educated at Michigan State College (Ph.D., chemistry, 1934). From 1962 he was Director of the Genetics Research Unit of the Carnegie Institution of Washington at Cold Spring Harbor. In 1952 he demonstrated, with Martha Chase, that only the DNA moiety of bacteriophage enters the host cell, and thus is the active component. He was awarded the Nobel Prize, with Delbrück and Luria, in 1969 for 'discoveries concerning the replication mechanism and the genetic structure of viruses'.

Hertwig, Karl Heinrich 1798–1881
Having studied medicine at Breslau, Hertwig travelled to Vienna, Munich and Berlin to educate himself further in veterinary medicine. In the late 1820s he carried out a series of classic studies on the transmission of rabies.

Hirst, George K. 1909–
Born in 1909, Hirst graduated from the Yale University Medical School in 1933. Much of his early work was done at the Rockefeller Institute Hospital, and it was in the laboratories of the International Health Division of the Rockefeller Foundation that Hirst made his observations of haemagglutination in 1941. Founder and architect of *Virology*, he guided the journal through its first 21 years as Editor-in-Chief from 1955–76.

Högyes, Andreas 1847–1906
Educated at the University of Budapest, Högyes held the Chair of General Pathology and Therapeutics there from 1883. In 1887 he served on a committee investigating Pasteur's anti-rabies treatment, and afterwards took part in the founding of an institute for rabies treatment in Hungary. Högyes wrote on rabies in contemporary textbooks.

Horsley, Sir Victor [A. H.] (knighted 1902) 1857–1916 FRS 1886
English surgeon who served as Secretary to the Government Commission appointed to evaluate Pasteur's anti-rabies treatment in 1885. At the time Superintendent at the Brown Institution, Horsley had clinical experience of the disease (one of his laboratory attendants succumbed to rabies) and was much impressed by Pasteur's work and ideas. Horsley's rabies research formed the basis for the legislation (muzzling and quarantining of imports orders) which freed Britain from rabies in the early twentieth century, until it was briefly reintroduced from 1918 to 1922.

Hufeland, Christoph Wilhelm 1762–1836
German pathologist, educated at the Universities of Jena and Göttingen, and later Professor of Pathology at Jena and Berlin. He was the author of a book on longevity, and founder and editor of *Hufeland's Journal der praktischen Arzneykunde*. He carried out research on smallpox vaccination.

Hunger, Friedrich Wilhelm Tobias 1874–1952
Born in Amsterdam, Hunger studied botany at the Universities of Leiden, Jena and Brussels. He investigated tobacco mosaic disease while stationed at Buitenzorg. With the exception of a year teaching tropical botany at the University of Utrecht (1905–06), Hunger stayed in Java until 1910. Later he turned to the history of botany and medicine, publishing two volumes on the life and work of Carolus Clusius, and becoming Director of the Institute for the History of Medicine and Science at Leiden.
(*Ber. dt. bot. ges.*, 1955, **68a** (Nachrufe): 34–7.)

Biographical notes

Isaacs, Alick   1921–67   FRS 1966

Having qualified in medicine at the University of Glasgow, Isaacs worked on influenza viruses under Burnet at the Walter and Eliza Hall Institute in Melbourne from 1948 to 1951, developing a strong interest in interference phenomena. Back in London he continued his studies at the National Institute for Medical Research, where with J. Lindenmann he discovered interferon in 1956.

(*Biogr. Mem. Fellows R. Soc.*, 1967, **13**: 205–22.)

Ivanovski, Dmitri Iosifovich   1864–1920

Born near St Petersburg, Ivanovski studied botany and plant physiology at the University of St Petersburg. Beginning in the late 1880s, he investigated tobacco diseases on Russian plantations, and published the first report of the filterability of the virus of tobacco mosaic disease in 1892, although his interpretation of his results was, and continued to be, wrong. From 1903, Ivanovski held the Chair of Plant Anatomy and Physiology at the University of Warsaw, and his research and teaching was largely concerned with plant physiology, on which subject he wrote a textbook.

(*Bact. Rev.*, 1972, **36**: 135–45.)

Jacob, François   1920–   For.M.R.S. 1973

Born at Nancy in 1920, Jacob intended to become a surgeon and studied medicine in Paris, but after the outbreak of war he joined the Free French Forces in London in 1940. He was wounded in North Africa, and later, and more severely, in Normandy in 1944. After the war Jacob completed his medical studies in Paris in 1947. Unable to practise surgery because of his war injuries, he turned to biological research and in 1950 joined the Institut Pasteur under Lwoff. He has made major contributions to the study of genetic mechanisms of bacteria and bacteriophages and was awarded the Nobel Prize, jointly with Lwoff and Monod, in 1965 for 'discoveries concerning genetic control of enzyme and virus synthesis'.

Joubert, Jules   1834–1910

French physicist, who was teaching physics in Paris when in 1876 he became associated with Pasteur's work on anthrax and chicken cholera. In physics, his interest was in research on alternating currents.

Johnson, James   1886–1952

American plant pathologist, educated at the University of Wisconsin, and remaining there throughout an active career, from 1922 as Professor of Horticulture, later of Plant Pathology. Johnson devoted his life's work to the improvement of the decreasing yields of tobacco in his home state of Wisconsin, studying diseases of the tobacco plant as well as aspects of nutrition and resistance.

(*Phytopathology*, 1954, **44**: 335–6.)

Keber, Gotthard August Ferdinand 1816–71
  German physician, educated at the Universities of Königsberg and Berlin. Rising to the rank of 'Regierungs-Medicinalrath' by 1858, Keber worked for many years in Danzig. His main areas of interest were the anatomy and physiology of molluscs, their nerve systems and their reproductive physiology, but he also carried out early filtration experiments on vaccinia lymph, in the course of microscopical studies.

Klebs (Theodor Albrecht), Edwin 1834–1913
  German pathologist who early in his career settled in Switzerland, but continued to move around a number of universities in Germany and Switzerland. On a personal level a difficult man, Klebs was a great innovator and a prolific writer, and was involved in many of the great advances in bacteriology in the latter half of the nineteenth century, although the truly great discoveries escaped him. He contributed to the early development of bacteria-proof filters, and wrote on rinderpest and vaccinia.

Kling, Carl 1887–1967
  Born in a small town in central Sweden and having completed his medical training in Stockholm, Kling, at the age of 24, as a young assistant in the Bacteriological Institute, was introduced to the problem of severe epidemics of poliomyelitis during the 1911 outbreak in Sweden. With his fellow investigators he made the most of the opportunity, and their study of the epidemic and the lessons to be learnt from it remains a model of its kind, although much of their work was not fully appreciated until much later.

Koch, Robert 1843–1910
  Together with Pasteur, Koch laid the foundations on which bacteriology, and ultimately virology, was built, and trained and inspired a number of the early workers in filterable virus research, notably Loeffler and Frosch who first demonstrated the filterability of foot-and-mouth disease virus. Around the turn of the century, Koch travelled widely in Africa, attempting among other things to develop a vaccine against rinderpest. He was awarded the Nobel Prize in 1905.

Kunkel, Louis Otto 1884–1960
  Born in Mexico, educated in the American mid-West, Kunkel gained a Ph.D. at Columbia University in 1915. From 1923 to 1932 he was at the Boyce Thompson Institute, and then accepted the directorship of the newly established Rockefeller Division for Plant Pathology at Princeton. Both at the Boyce Thompson and at the Rockefeller Kunkel carried out fundamental work on insect transmission and other aspects of the 'yellows' type of plant virus diseases.
  (*Biogr. Mem. natn. Acad. Sci.*, 1965, **38**: 145–60.)

Laidlaw, Sir Patrick [Playfair] (P. P. Laidlaw, knighted 1935) 1881–1940 FRS 1927
  Educated at Cambridge and Guy's Hospital Medical School, Laidlaw in 1922 was put in charge of a new research programme into filterable virus diseases

at the National Institute for Medical Research, established at Hampstead two years before. Beginning with work on dog distemper, the group went from strength to strength, culminating in the isolation of influenza virus in ferrets in 1933, and subsequent adaptation of the virus to growth in mice.

(*Obit. Not. Fell. R. Soc. Lond.*, 1939–41, **3**: 427–47.)

Landsteiner, Karl   1868–1943   For M.R.S. 1941
Graduating from the Medical school of the University of Vienna in 1891, Landsteiner spent the next five years studying chemistry in Zürich, Würzburg (with Emil Fischer) and Munich, before returning to Vienna as assistant to Max Gruber. In Vienna, before the outbreak of World War I, Landsteiner made notable contributions to virus research, especially concerning the virus of poliomyelitis, before devoting himself fully to immunology and emigrating to the United States in 1919. He was awarded the Nobel Prize for 'his discovery of the human blood groups' in 1930.

(*Obit. Not. Fell. R. Soc. Lond.*, 1945–48, **5**: 295–324.)

Lazear, Jesse William   1866–1900
Lazear graduated in arts at Johns Hopkins University in 1889, and in medicine at Columbia University 1892. During a year in Europe he spent some time at the Institut Pasteur in Paris. A member of the American Yellow Fever Commission, he died of yellow fever in Cuba on 28 September 1900.

Ledingham, Sir John [Charles Grant] (J. C. G. Ledingham, knighted 1937) 1875–1944   FRS 1921
Born in Banffshire and educated in Aberdeen, Ledingham was Director of the Lister Institute from 1931 to 1943, and Professor of Bacteriology in the University of London. Fowl pox and vaccinia were among his research interests; in 1929 at a Royal Society discussion he stated his belief that not only were viruses particulate, but that the larger ones could be seen readily with the microscope.

(*Obit. Not. Fell. R. Soc. Lond.*, 1945–48, **5**: 325–40.)

Levaditi, Constantin   1874–1953
Romanian by birth and dedicated to research from his earliest days in medicine, Levaditi worked for a time for Babes but was determined to leave for Paris where he arrived in 1898. At first he worked in the laboratory of the Hôtel-Dieu, and later at the Collège de France. He spent a year with Ehrlich at Frankfurt 1899–1900. Returning to Paris he entered Metchnikoff's laboratory at the Institut Pasteur where he stayed. Since his stay with Ehrlich, Levaditi was interested above all in syphilis, but he also undertook, with Borrel, a systematic study of pathogenic viruses. With Landsteiner he demonstrated the filterability of the virus of poliomyelitis, and later studied this virus together with Kling and Lepine. Also in the 1920s he did research on encephalitis lethargica and rabies.

(*Annls Inst. Pasteur, Paris*, 1953, **85**: 535–9.)

Biographical notes

Lindenmann, Jean   1924–
Swiss medical microbiologist, born in Zagreb of Swiss parents. Lindenmann graduated M.D. from the University of Zürich in 1951, and from 1956 to 1957 worked on interference phenomena at the National Institute for Medical Research in London with Alick Isaacs, the work which culminated in the discovery of interferon. More recently, as professor at the Institute for Medical Microbiology in the University of Zürich, Lindenmann has been studying aspects of immunogenic mechanisms.

Lipschütz, Benjamin   1878–1931
Austrian pathologist and dermatologist. Having graduated at Vienna in 1902, Lipschütz worked at the Institute of Pathology there before going to Paris to study cytology at the Institut Pasteur and dermatology at the Hôpital St-Louis. Having qualified in dermatology in Vienna he became head of the Dermatology Department of the Children's Hospital. Lipschütz is remembered for his contributions to the study of virus diseases characterized by the forming of inclusion bodies, for coining the term '*strongyloplasma*' as an alternative to von Prowazek's '*chlamydozoa*', and above all for his classic studies of herpesvirus diseases.

Lode, Alois   1866–1950
Born and educated in Vienna, Lode began his career as assistant to Max Gruber in the Institute of Hygiene. In 1899, Lode moved to Innsbruck, where he became Professor of Hygiene in 1908, and Director of the Institute of Hygiene. With F. Gruber he reported, in 1901, the occurrence of fowl plague in Innsbruck, carried over the Alps with the stock of an itinerant poultry merchant, a couple of months after the first outbreak in Ferrara reported by Centanni.

Loeffler, Friedrich   1852–1915
Loeffler studied medicine in Würzburg until interrupted by service in the Franco-Prussian war of 1870–71, after which he continued training in the Army Medical Institutes in Berlin from which he graduated in 1874. From 1879, Loeffler was seconded to the Kaiserliche Gesundheitsamt in Berlin, where he was soon to be joined by Robert Koch. In 1888, Loeffler was appointed to the Chair of Hygiene at Greifswald, from which he returned to Berlin only two years before his death, as Director of the Institut für Infektionskrankheiten, in 1913. Loeffler made many important contributions to the field of pathology, but in the history of virology it is his study, with Paul Frosch, of the pathogen of foot-and-mouth disease which is best remembered.
(*Zentbl. Bakt. ParasitKde*, Abt. I, Orig., 1932, **125**: i–xx.)

Luria, Salvador E.   1912–
Born in 1912 and educated in Turin, Luria was working as a Research Fellow in the Radium Institute in Paris when World War II broke out. At the fall of France in 1940 he escaped to the United States, where he soon met Max Delbrück and discovered their common objectives. Together Luria and Delbrück laid the foundations for the American school of molecular genetics

as it evolved in the 1940s and 1950s. Since 1959 Luria has held a number of Chairs at Massachusetts Institute of Technology, from 1970 as Institute Professor in the Department of Biology. With Delbrück and Hershey, Luria was awarded the 1969 Nobel Prize for 'discoveries concerning the replication mechanism and the genetic structure of viruses'.

Lwoff, André M. 1902–

Born in 1902, Lwoff joined the Institut Pasteur at the age of 19, having graduated in science and done one year of medical studies which he subsequently completed while working in the laboratory. Much of his early research was concerned with the development and morphogenesis of parasitic ciliates and the nutrition of protozoa. In the late 1940s Lwoff took up the problem of lysogeny, and by his studies of the induction phenomenon contributed greatly to the understanding of genetic processes in bacteriophage. Awarded the Nobel Prize, jointly with Jacob and Monod, in 1965 for 'discoveries concerning genetic control of enzyme and virus synthesis'.

M'Fadyean, Sir John (knighted 1905) 1853–1941

The son of a Wigtown farmer, M'Fadyean at first chose to work with his father on the land rather than become a solicitor; then, in the face of increasing difficulties for farming, he decided to study veterinary medicine. Having graduated from Dick (later Royal) Veterinary College at Edinburgh, where he was taught physiology by McKendrick, M'Fadyean returned there only six months later to teach anatomy, while at the same time studying science and medicine at Edinburgh University, qualifying in medicine and surgery in 1882 and in science in the following year. He immersed himself enthusiastically in what could be called the Golden Age of Pathology; in 1888 he founded, and until his death edited, the *Journal of Comparative Pathology and Therapeutics*, and in 1892 he became Dean (later Principal) of the Royal Veterinary College of London, and Professor of Pathology and Bacteriology. An outstanding personality both as teacher and researcher, M'Fadyean demonstrated the filterability of the pathogen of African horse sickness in January 1900.

McKendrick, John Gray 1841–1926 FRS 1884, FRSE 1873

Born in Aberdeen, of celtic stock, McKendrick was orphaned as an infant and brought up in relative poverty by friends and relations. At 13 he became a 'herd laddie', then was apprenticed to a solicitor where he served for six years with a legal firm in Aberdeen. During this time he was taken up by various 'cultured families' in the town, who introduced him to music, poetry, astronomy and even the use of the microscope – at that time still a favourite drawing-room diversion in Britain. His interest thus awakened, McKendrick began studying science, and eventually medicine, outside office hours. In spite of difficulties and occasional setbacks, he qualified in medicine at the University of Aberdeen in 1864. A gifted teacher and 'observer of physiological facts', he taught physiology at the University of Edinburgh and, from 1876 until he retired in 1906, at the University of Glasgow.

(*Proc. R. Soc. Lond.* B, 1926, **100**: xiv–xviii.)

Biographical notes

von Magnus, Preben   1912–73

Born and educated in Copenhagen, von Magnus worked at the Danish State Serum Institute throughout his career, from 1959 as its Director. With his wife, Dr Herdis von Magnus, he made considerable contributions to poliomyelitis research. In 1951 he recorded the propensity of influenza virus in certain circumstances to produce progeny with reduced infectivity and deficient in RNA, which has since been known as the 'von Magnus phenomenon'.

(*Lancet*, 1974, **i**: 35.)

Maitland, Hugh Bethune   1895–1972

Born in Ontario, educated at the University of Toronto, Maitland taught bacteriology there until, after a year in Germany, he joined the Lister Institute as bacteriologist to the commission on foot-and-mouth disease set up by the Ministry of Agriculture. Here, with S. P. Bedson and Yvonne Burbury, he made important contributions to the study of foot-and-mouth disease virus. In 1927, Maitland was appointed to the Chair of Bacteriology at the University of Manchester where, with his wife, he succeeded in growing vaccinia virus in a simple tissue culture system subsequently known as Maitland's medium; a modification of this system was used many years later for growing poliomyelitis virus for use in the preparation of Salk vaccine.

(*J. med. Microbiol.*, 1973, **6**: 253–8.)

Mayer, Adolf   1843–? (still alive in late 1930s)

Born in Germany and educated at the Universities of Heidelberg, Ghent and Halle, Mayer became Professor of Chemical Technology at the University of Heidelberg in 1875 (during this period he wrote papers on a variety of subjects, including alcohol fermentation, street cleaning, artificial butter, burning quality of tobacco, etc.). In 1876, Mayer moved to Holland as Director of the Agricultural Experiment Station at Wageningen, where in 1886 he conducted an official investigation of the problem presented by tobacco mosaic disease. In 1904 he returned to Heidelberg University as Professor of Botany, but also devoting time to political economy and the writing of drama in poetry and prose, long outliving all his contemporaries (he was still alive shortly before the outbreak of World War II) – even Beijerinck, who was 80 when he died, and who was his junior by eight years.

Monod, Jaques Lucien   1910–76   For.M.R.S. 1968

Having graduated in natural sciences in Paris in 1931, Monod taught and studied protozoa, then proceeded to develop standard quantitative methods for the growth of bacterial cultures. By the time his thesis on this subject was published, Monod was no longer to be found in the laboratory; he had joined the French Resistance. After the war he joined Lwoff's department at the Institut Pasteur, and turned his attention to the formation of specific enzyme proteins. In 1958, Monod began his collaboration with Jacob which led them to the formation of the concept of messenger RNA. Monod shared the 1965 Nobel Prize with Jacob and Lwoff for 'discoveries concerning genetic control of enzyme and virus synthesis'.

(*Biogr. Mem. Fellows R. Soc.*, 1977, **23**: 385–413.)

Biographical notes

Montagu, Lady Mary Wortley   1689–1762
A beauty blemished by an early attack of smallpox which had left her with a grudge against the medical profession, Lady Mary discovered inoculation as practised in Turkey when in 1716 she went with her husband to Constantinople upon his appointment as Ambassador. She had her own small son successfully inoculated in Constantinople, and after her return to Britain she campaigned for the acceptance of the practice in Britain and on the continent where she later lived for protracted periods. Her early impressions of the practice of inoculation in Turkey are well described in her letters, first published in 1763.
   (*J. Hist. Med.*, 1953, **8**: 390–405.)

Murphy, James Bumgardner   1884–1950
Having graduated from the Medical School of Johns Hopkins University, Murphy joined Peyton Rous at the Rockefeller Institute in 1911, where they developed the method for growing Rous' sarcoma virus in embryonated eggs. This work furthered his interest in inter-species grafts of mammalian tumour tissue, and in factors determining tissue specificity.
   (*Biogr. Mem. natn. Acad. Sci.*, 1960, **34**: 183–203.)

Negri, Adelchi   1876–1912
Born in Perugia, Negri graduated from the University of Pavia in 1900 and became assistant to Golgi in the Institute of Pathology, and it was here he carried out his studies on rabies and described the inclusion bodies named after him, which he insisted were protozoal parasites. In 1905 he demonstrated the filterability of vaccinia virus. From 1908 he was Professor of Microbiology in the University of Pavia, and for the last three years of his life he mounted an active campaign against malaria in the Lombardy region.
   (*Parasitology*, 1912, **5**: 151–4.)

Nicolle, Charles [J. H.]   1866–1936
The son of a physician in Rouen, Nicolle qualified in medicine in Paris in 1893, having in 1892 followed, on his father's advice, the newly established course in microbiology at the Institut Pasteur. After a few years of teaching and hospital practice in his native Rouen, Nicolle accepted the post of Director of the Institut Pasteur in Tunis, where he spent the rest of his life. He made contributions to the study of a number of virus diseases, notably rinderpest and measles, but is best known for his work on the transmission of typhus fever for which he was awarded the Nobel Prize in 1928.
   (*Lancet*, 1936, **i**: 560; *Annls Inst. Pasteur, Paris*, 1936, **56**: 353–8.)

Nocard, Edmond I. E.   1850–1903
Educated at the Veterinary School of Alfort, near Paris, Nocard eventually became its Director in 1889. He was a member of the team which obtained the interesting, but for virus research in general confusing, results with bovine pleuropneumonia in 1898. Nocard also studied a number of true virus diseases such as rabies, foot-and-mouth disease and sheep pox.
   (*Lancet*, 1903, **ii**: 430; 781.)

Biographical notes

Noguchi, Hideyo (Seisaka) 1876–1928

Born in extreme poverty in a tiny Japanese village, Noguchi graduated from the Imperial University of Tokyo in 1897. At the turn of the century he came to the United States to work with Simon Flexner at the University of Pennsylvania, and in 1904 joined the Rockefeller Institute under Flexner's directorship. Noguchi was a tireless and enthusiastic worker in several different areas of bacteriology, and contributed to the study of rabies and of yellow fever. A colourful and occasionally controversial personality (his work on trachoma and yellow fever, especially, gave rise to heated argument), he changed his given name of Seisaka to Hideyo which means 'great-man-of-the-world'. He died of yellow fever in the course of his researches in Africa.

(*Bull. Hist. Med.*, 1959, **33**: 1–20.)

Olitsky, Peter K. 1886–1964

Born in New York and educated at the University of Cornell, where he graduated M.D. in 1909, Olitsky came to the Rockefeller Institute in 1917, where he continued to pursue his interest in viruses, and virus diseases throughout the rest of his career. Much of Olitsky's work was on poliomyelitis (with Sabin), and on the encephalitides of man and the lower animals, although he made an isolated attempt to cultivate the virus of tobacco mosaic disease in 1925.

(*Lancet*, 1964, **ii**: 423.)

Panum, Peter Ludwig 1820–85

Danish physician and physiologist. Born on the island of Bornholm, Panum went to school in the border town of Flensborg where his father, a surgeon in the armed forces, was stationed. He began his university education in Kiel, but soon transferred to the University of Copenhagen. Qualifying at the age of 25, Panum was sent, the following year, to the Faeroes to investigate a severe outbreak of measles in this isolated community. The report he produced has become a landmark in the history of epidemiological studies. Later, his interest turned towards physiological chemistry and experimental physiology, and he spent some time with Claude Bernard in Paris. As Professor of Physiology at the University of Copenhagen from 1864, Panum introduced the study of experimental medicine and physiology in the face of stiff opposition from the rest of the Faculty of Medicine.

Paschen, Enrique 1860–1936

Born in Mexico (his father was a businessman and German consul there), Paschen was educated at the University of Heidelberg and in 1890 became assistant in the State Inoculation Department in Hamburg, beginning an association which was to last until he retired. The cholera epidemic of 1892 and the 1914–18 war saw him back in clinical work, but his main interest was always smallpox research, and he travelled repeatedly to study outbreaks abroad. In 1906, Paschen described the elementary bodies named after him.

(*Dt. med. Wschr.*, 1936, **62** (ii): 1891–2.)

Biographical notes

Pasteur, Louis 1822–95 For.M.R.S. 1869
French chemist turned microbiologist and, with Robert Koch, the main protagonist of the rising science of bacteriology in the latter part of the nineteenth century. The width spanned by Pasteur's researches is astonishing, and his discoveries are legion, from his early (1848) observations of the two optically active forms of tartaric acid, via fermentation studies and diseases of the silkworm, to the epoch-making work on vaccines against fowl cholera, anthrax and finally rabies, in the 1880s.

Paul, John Rodman 1893–1971
Educated at Princeton, then the Medical School of the Johns Hopkins University, Paul served for a year as bacteriological technician at a United States Base Hospital in France in 1917–18. From 1928 he was on the Faculty of the Medical School at Yale University, where his concept of 'clinical epidemiology' developed throughout the 1930s. During the 1940s and 1950s Paul made important contributions to the study of poliomyelitis. When cardiac problems forced his retirement from the WHO Serum Bank in 1966, he devoted his remaining years to scholarly activities within the Yale Department of the History of Science and Medicine, culminating in the publication, a few months before his death, of *A history of poliomyelitis.*
(*Biogr. Mem. natn. Acad. Sci.*, 1975, **47**: 323–68.)

Pirie, Norman Wingate 1907– FRS 1949
Born in 1907, Pirie was educated at Cambridge, where he first turned to biochemistry, and continued to revel in the subject, in the stimulating atmosphere of Gowland Hopkins' laboratory. His famous partnership with F. C. Bawden began in the 1930s, and together they made contributions of fundamental importance to the study of the nature of plant viruses. In 1940, Pirie joined Bawden at Rothamsted Experimental Station, becoming head of its Biochemistry Department in 1947. In later years his attention has been focused mainly on nutritional and population problems.

von Prowazek, Stanislaus J. M. 1875–1915
Born in what was then Bohemia and is now Czechoslovakia, von Prowazek studied natural history and philosophy first at the University of Prague, then at Vienna. He possessed a marked talent for meticulous observation and very soon became an enthusiastic microscopist, and an outstanding one. After research at the Zoological Institute in Vienna and at the Research Station of Trieste he spent a year with Ehrlich in Frankfurt and another one with Hertwig at Munich. In 1906, von Prowazek became Director of the Institut für Schiffs- und Tropenkrankheiten in Hamburg. In 1910–12 he travelled in Samoa and Sumatra studying trachoma. In the summer of 1913 he made a short study of typhus fever in Serbia, and in 1914 a similar one, with da Roche-Lima, in Constantinople – studies which made him the natural choice for the Ministry of War to send to investigate an outbreak of typhus fever in a prisoner-of-war camp in Cottbus, northeast of Dresden. Von Prowazek arrived in Cottbus just before Christmas 1914. On 17 February, he died there

of typhus fever. He is remembered in the history of virus research for his observations of inclusion bodies and his formulation of the chlamydozoa concept.

(*Arch. Protistenk.*, 1915–16, **36**: i–xii.)

Pylarino, Giacomo   1659–1718
   Born on the Greek island of Cephalonia, Pylarino first studied law in Venice, and afterwards medicine at Padua. He travelled to Constantinople and Syria, practised medicine for a while in Aleppo, and finally via Egypt reached Smyrna, where he settled, attached to the Venetian Embassy. He eventually returned to Venice, where he published *Nova et tuta variolas excitandi per transplantationem methodus,* an account of the method of inoculating smallpox as he had developed it on the basis of the practice of the Greek witch women.

Rayleigh, Lord, John William Strutt, 3rd Baron Rayleigh   1842–1919   FRS 1873   PRS 1908
   English mathematician and physicist who succeeded Clerk Maxwell in the Cavendish Chair which was re-established specially for him after Clerk Maxwell's death in 1879. By 1884, Rayleigh was finding the task too onerous and resigned, to retire to his private laboratory on his Essex estate (2845 ha), although he was later professor at the Royal Institution from 1887 to 1905. He made a number of important contributions to the study of sound and of light scattering and optics; he was awarded the 1904 Nobel Prize for physics for his discovery of argon.

Reed, Walter   1851–1902
   American army surgeon, educated at the University of Virginia and the Bellevue Hospital Medical College. In 1898, Reed investigated the epidemiology of typhoid fever in army camps, and in 1900 he was appointed head of the army's Yellow Fever Commission in Havana. Between 1900 and 1902 the Commission carried out a controlled study, using human volunteers, of the nature and transmission of the disease. The findings of the Commission, that yellow fever is caused by a filterable virus carried by *Aëdes aegypti,* formed the basis for the subsequent eradication of the disease in Cuba and the Canal Zone. Reed died after his return to the United States, not of yellow fever but of peritonitis, following appendicectomy.

Remlinger, Paul [Ambroise]   1871–1964
   As Director of the Institut Pasteur in Tangier, Remlinger long outlived all the other *vieux pasteuriens* who had been at the Paris institute in the time of Pasteur himself. Living the life of a near recluse in the institute in Tangiers, he studied and wrote on rabies for more than 50 years, totally dedicated to his research.

(*Annls Inst. Pasteur, Paris,* 1965, **108**: 689–94.)

Rivers, Thomas Milton   1888–1962
   American virologist. Rivers grew up on a farm in Georgia and graduated in medicine at Johns Hopkins University in 1915, after his studies had been interrupted by what was diagnosed as progressive muscular atrophy. Rivers

surprisingly overcame this by 18 months of hard work as a medical technician in Panama. In 1922 he was given the opportunity to develop laboratories for virus research at the Rockefeller Institute Hospital, where he at first studied varicella with Tillett, only to discover that they were actually dealing with the so-called virus III, a hitherto unknown latent virus of the rabbit. Later, Rivers worked on psittacosis and post-vaccinial encephalitis, and carried out clinical studies of louping ill, Rift Valley fever and lymphocytic choriomeningitis. In 1928 he edited the first edition of the classic volume *Filterable viruses.*

(*Biogr. Mem. natn. Acad. Sci.*, 1965, **38**: 263–94.)

Rivolta, Sebastiano   1832–93
   Italian pathologist, graduated in veterinary medicine, taught pathology and pathological anatomy. From 1868 he was professor at the University of Pisa. He retired early due to illness and wrote *Dei parassiti vegetali*, 1873, and, with Delprato, *L'ornitoitria o la medicina degli uccelli domestici e semidomestici*, 1880. He observed fowl pox inclusions as early as 1869.

Rous, Francis Peyton   1879–1970   For.M.R.S. 1940
   Educated at Johns Hopkins University and Medical School. Shortly after joining the Rockefeller Institute, Rous was given the opportunity, in 1909, to investigate a spindle-celled sarcoma on a Plymouth Rock hen brought to the institute for examination. Rous was able to propagate the tumour in series in related chickens, by injecting minced tumour tissue, and shortly afterwards, to transmit the sarcoma, which came to bear his name, by means of cell-free filtrates. In spite of much negative reaction to his discovery ranging from indifference to outspoken opposition, Rous never became embittered or discouraged, but continued his research in this and other fields throughout a long and active career. Fortunately he lived long enough to enjoy vindication of his theories, and to be given full recognition. Assisting Flexner in editing the *Journal of Experimental Medicine* since his early days at the Rockefeller, Rous was appointed co-editor in 1921, and served as editor of this journal for nearly 50 years. Rous was finally awarded the Nobel Prize, long overdue in his case, in 1966 for his 'discovery of tumour-inducing viruses'.

(*Biogr. Mem. natn. Acad. Sci.*, 1976, **48**: 275–306; *Biogr. Mem. Fellows R. Soc.*, 1971, **17**: 643–62.)

Roux [Pierre Paul], Emile   1853–1933   For.M.R.S. 1913
   Having lost his father early, Roux was brought up and educated in relative poverty in an academic environment; it bred in him dedication to science and ascetic attitudes – even when Director of the Institut Pasteur Roux insisted on living in rooms at the top of the institute normally occupied by junior members of staff, on a salary to match. Roux began his medical studies at Clermont-Ferrand where he was influenced by Duclaux, and completed them in Paris. In 1878, Duclaux introduced him to Pasteur, who engaged him as assistant, allegedly because Pasteur, who found it difficult to inoculate guinea pigs, needed someone with Roux's medical expertise to carry out such operations. Roux was then, and for a long time to come, the only medically trained worker in Pasteur's laboratories, although all the others had passed through the rigours of the Ecole Normale. Throughout his association with Pasteur,

Roux played a major part in many of the great discoveries, as his name on the papers describing them shows. In 1904 he succeeded Duclaux as director of the Institut Pasteur in Paris.

(*Obit. Not. Fell. R. Soc. Lond.*, 1932–35, **1**: 197–204.)

Roux, Wilhelm   1850–1924
   Born in Jena, Roux was educated at the Universities of Jena, Berlin (where he studied under Virchow) and Strasbourg. The founder of the discipline of experimental embryology Roux was, from 1888, Director of the Institut für Entwicklungsgeschichte und Entwicklungsmechanik created for him in Breslau, and also founded and edited *Archiv für Entwicklungsmechanik*. From 1895 to 1921 he was professor at the University of Halle.

Salaman, Redcliffe Nathan   1874–1955   FRS 1935
   Born in London, Salaman was educated at Cambridge and the London Hospital, qualifying in 1901. Illness forced him to interrupt his medical career when it had hardly begun; after he recovered he decided to retire to the country and devote himself to a less strenuous career in plant pathology. Having the advantage of private means, Salaman was eventually able to establish the Potato Virus Research Station at Cambridge where much important work for Britain's potato crops was carried out, and where Kenneth Smith and F. C. Bawden gained valuable experience early in their careers.

(*Biogr. Mem. Fellows R. Soc.*, 1955, **1**: 239–45.)

Sanarelli, Guiseppe   1864–1940
   Italian pathologist. He graduated in medicine at the University of Siena in 1889, and in 1892 started a life-long association with the Institut Pasteur in Paris by spending a scholarship year with Metchnikoff. In the late 1890s, Sanarelli was for a while Director of the recently established Institute of Hygiene in Montevideo, where he isolated and described the virus of myxomatosis in 1898. The year before, Sanarelli had with somewhat surprising lack of concern used human subjects for experimental inoculation of cultures in an attempt to establish '*Bacillus icteroides*' as the pathogen of yellow fever. Despite the storm of protest which followed, Sanarelli described similar experiments in his paper on myxomatosis. After his return from Montevideo he was professor first at the University of Bologna, later at Rome.

(*Annls Inst. Pasteur, Paris*, 1940, **64**: 357–8.)

Sanderson, John Scott Burdon, see Burdon-Sanderson

Sanfelice, Francesco   1866–1945
   Born in Rome, educated at the University of Naples, Sanfelice held Chairs of Hygiene in the Universities of Cagliari, Messina, Modena, Bari and finally, from 1931, Pisa. A life-long interest in malignant tumours led him to study the tumour-like lesions of fowl pox. In the course of this study he demonstrated, in 1914, that the pathogen of fowl pox reacted to various forms of chemical and physical treatment as a nucleoproteid (sic).

(*Riv. ital. Ig.*, 1945, IV–V, n. 3–4: 3–15.)

Biographical notes

Schäfer, Werner 1912–
German virologist, born 1912, educated at Giessen. Dr. med vet 1938. After the war he joined the virus research group of the Kaiser Wilhelm Institut für Biochemie. This group eventually became the Max Planck Institut für Virusforschung in 1954, and he became (1956) head of the animal virus division. His work includes studies on the structure, chemistry and immunology of influenza viruses, and, more recently, of oncogenic RNA viruses. In 1955, Schäfer identified fowl plague virus as an avian influenza A virus.

Schramm, Gerhard Felix 1910–69
German biochemist. He was educated at the Universities of Göttingen and Munich, and for many years associated with the Kaiser Wilhelm, later the Max Planck, Institutes in Berlin and Tübingen. From 1956 he was head of the Biochemistry Department of the Max Planck Institute for Virus Research at Tübingen. In 1956 he isolated, with Gierer, and simultaneously with the independent work by Fraenkel-Conrat in Berkeley, the infectious nucleic acid of tobacco mosaic virus.

Schrödinger, Erwin 1887–1961 For.M.R.S. 1949
Educated in his native city of Vienna, Schrödinger stayed there until 1920 when he took up a post in Zürich, where he formulated his epoch-making wave equation in 1926. In 1927 he moved to Berlin; but by 1933 he was no longer able to stand the political climate of the Führer's Germany (Schrödinger was not Jewish and had in fact been raised in the Roman catholic faith in Vienna) and left the country, eventually living in Dublin from 1940 to 1956, when he returned to Vienna. In Dublin, Schrödinger wrote the slender but influential volume *What is life?*. He found it difficult to accept some of the new developments in theoretical physics formulated in Copenhagen and in Cambridge in the 1930s, and this may have changed the direction of his thinking. He was awarded the Nobel Prize in physics, jointly with P. A. M. Dirac, in 1933 for 'discovery of new fruitful forms of atomic theory'.
(*Biogr. Mem. Fellows R. Soc.*, 1961, **7**: 221–8.)

Shope, Richard E. 1901–66
Having obtained his M.D. from the University of Iowa in 1924, Shope began his long association with the Rockefeller Institute in 1925; in 1952 he became Professor of Animal Pathology. His main contributions have been in the areas of swine influenza and transmissible rabbit fibromas and papillomas.

Sigurdsson, Björn 1913–59
A graduate of the Medical School of the University of Reykjavik, Iceland, in 1937, Sigurdsson went straight into research at the Institute of Pathology in Reykjavik where diseases of sheep were traditionally a major problem to a nation heavily dependent on its sheep farming. Later, he spent some time in Copenhagen at the Carlsberg Laboratories with Albert Fischer, and at the State Serum Institute, and in the early 1940s he went to the Rockefeller Institute, where a congenial group included W. M. Stanley and R. E. Shope. In late 1943, Sigurdsson was able to return to Iceland with the promise of

Rockefeller support to build an Institute for Experimental Pathology at Keldur. Intrigued by the problem of transmission of scrapie since his early days in pathology, Sigurdsson recognized at Keldur (with still better electron microscopes at his disposal) the histopathological similarities between scrapie brains and certain human pathological lesions, for example in multiple sclerosis. He initiated the concept of the slow virus diseases, and pointed out its potential interest within comparative pathology, in a series of lectures in London in 1954, before he died at the age of 46 in 1959.

Smith, Kenneth Manley   1892–   FRS 1938
 Born in Scotland and educated at the Royal College of Science, Smith became director of the virus research unit of the Agricultural Research Council at Cambridge in 1939. In later years he turned his attention to insect viruses, a field in which he has been very active at home and abroad since his official retirement. Smith has written, and contributed to, a number of textbooks in the fields of entomology and plant and insect virology.

Smith, Wilson   1897–1965   FRS 1949
 Brought up by a widowed mother with a firm belief in education, Smith at the age of 18 joined the RAMC, serving in France and Belgium during World War I. Although for him, as for many of his contemporaries, this was a profoundly unhappy experience, it served him well in inducing him to embark on the study of medicine in 1919. Having graduated M.B., Ch.B. at Manchester and adding a diploma in bacteriology, he joined the National Institute for Medical Research in 1927, when work on filterable viruses was becoming established in a group which included P. P. Laidlaw, W. E. Gye, J. E. Barnard, W. J. Elford and C. H. Andrewes. Smith made important contributions to the study of influenza in the early 1930s with Laidlaw and Andrewes, and later with C. H. Stuart-Harris.
 (*Biogr. Mem. Fellows R. Soc.*, 1966, **12**: 479–87.)

Stanley, Wendell Meredith   1904–71
 American biochemist. He joined the then recently established Rockefeller plant pathology division at Princeton under Kunkel in 1931, straight from a postgraduate year in Germany, to work on tobacco mosaic virus. In 1948 he became the first director of the Laboratory for Virus Research established for him on the Berkeley campus of the University of California. He shared the Nobel Prize for chemistry in 1946 with J. B. Sumner and J. H. Northrop for 'preparation of virus protein in a pure form'.
 (*Nature, Lond.*, 1971, **233**: 149–50.)

Staudinger, Hermann   1881–1965
 German chemist. He was educated at the Universities of Halle, Munich and Darmstadt. He worked in Zürich from 1912 to 1926, and afterwards at the University of Freiburg im Breisgau until he retired in 1951. In the 1920s, Staudinger brought in the then revolutionary concept of macromolecules which for many years was resisted by the majority of eminent chemists, until advances in X-ray crystallography confirmed his predictions; his work on polymers became the foundation for the rise of the plastic industry. Staudinger

received the 1953 Nobel Prize in chemistry for 'discoveries in the field of macromolecular chemistry'.

Stent, Gunther S. 1924–
    Born in Berlin in 1924, Stent obtained a Ph.D. in physical chemistry from the University of Illinois in 1948, and went to the California Institute of Technology just when exciting discoveries were being made in the field of molecular biology and genetics. After a couple of years abroad, in Copenhagen and at the Institut Pasteur in Paris, he returned to an appointment at the University of California at Berkeley where he has remained, since 1959 as Professor of Bacteriology and Virology. In 1963 he published *Molecular biology of bacterial viruses*; he has also contributed many essays on the development of the discipline of molecular biology and genetics.

Storey, Harold Haydon 1895–1969 FRS 1946
    English plant pathologist. Educated at Cambridge, Storey spent most of his working life in East Africa, where he demonstrated the viral aetiology of a number of diseases attacking important East African crop plants, and developed a technique for insect injection which proved *inter alia* the aphid transmission of 'rosette' disease of groundnuts.
    (*Biogr. Mem. Fellows R. Soc.*, 1969, **15**: 239–46.)

Sumner, James Batcheller 1887–1955
    Having graduated from Harvard in organic chemistry in 1910, Sumner spent some time travelling in Europe, and in 1914 began a career in teaching and research at Cornell University. In 1926 he succeeded in preparing crystals of urease, the first ever crystallization of an enzyme. For this work Sumner shared the 1946 Nobel Prize in chemistry with J. H. Northrop and W. M. Stanley.
    (*Biogr. Mem. natn. Acad. Sci.*, 1958, **31**: 376–96.)

Svedberg, The(Theodor) 1884–1971 For.M.R.S. 1944
    Having just graduated from the University of Uppsala, Svedberg in 1905 read Zsigmondy's recently published *Zur Erkenntnis der Kolloide* and became committed to the study of colloidal systems as a means of elucidating processes in living tissues. In 1926, Svedberg, with Robin Fåhraeus, developed the first ultracentrifuge, which was gradually perfected over the following decade. He was awarded the Nobel Prize in chemistry in 1926 for 'work on disperse systems'.
    (*Biogr. Mem. Fellows R. Soc.*, 1973, **19**: 19–63.)

Theiler, Sir Arnold 1867–1936
    Born and educated in veterinary medicine in Switzerland, Theiler was attracted early by the promise of adventure in Africa, and started a career in private veterinary practice in Johannesburg in 1891. Aware of the challenge of attempts to control the epizootics which were continually sweeping the African continent, he became official veterinarian of Transvaal in charge of controlling the rinderpest outbreak of 1896. Having served as veterinary officer with the Boer forces in the Boer War, Theiler subsequently was appointed

veterinary bacteriologist of Transvaal by the new British administration. He made many contributions to the control of infectious disease in domestic animals in Africa; he was also interested in improving veterinary education, and campaigned tirelessly for support to research.

(*Mod. vet. Pract.*, 1959, **40** (19): 46–7.)

Theiler, Max 1899–1972
Born in Pretoria, son of the above, Theiler received his medical education at the Universities of Basle and Capetown, and afterwards St Thomas' Hospital and the London School of Tropical Medicine and Hygiene. Later he joined the Department of Tropical Medicine at Harvard Medical School, and in 1930 came to the Rockefeller Foundation laboratories, where he succeeded in transmitting yellow fever to laboratory mice, and later in developing an attenuated live virus vaccine against yellow fever. In 1951, Theiler was awarded the Nobel Prize for 'discoveries concerning yellow fever and how to combat it'.

Timoni(s), Emanuele ?–1718
Timonis was a physician practising in Constantinople, where he may have been physician to the household of Edward Wortley Montague, which included his wife Lady Mary Wortley Montague and their small son who was inoculated with smallpox during their stay. Timonis had studied medicine at the Universities of Padua and Oxford; having known Pylarino in Smyrna, Timonis in 1713 communicated a letter on the practice of variolation to the Royal Society in London (printed in the *Transactions*, 1714).

Twort, Frederick William 1877–1950 FRS 1929
The son of a doctor, Twort followed the family tradition (as did his brother and sometime collaborator, Charles Claud Twort), qualifying at St Thomas' Hospital Medical School in 1900. After a few years of assisting William Bulloch in the diagnostic laboratories of the London Hospital, Twort became Superintendent of the Brown Institution in 1909, working in comparative seclusion for 35 years until the institute was destroyed by enemy action in 1944. While working on Johne's bacillus, Twort became interested in filterable viruses, and in 1915 discovered the lytic phenomenon which d'Herelle later named bacteriophagy.

(*Obit. Not. Fell. R. Soc. Lond.*, 1950–51, **7**: 505–17.)

Vida, Marco Girolamo *c.* 1490–1566
An Italian poet who became bishop of Alba in 1532, and as such was a member of the Council of Trent, where he may have known Fracastoro. He was commissioned by Pope Leo X to write the Christian epic the *Christiad* (1535). Vida is probably better known for his *Ars poetica*, and the didactic poems on the silkworm and on chess.

Vinson, Carl George 1891–1964
An American biochemist who in the late 1920s at the Boyce Thompson Institute developed methods for precipitation of tobacco mosaic virus, later used by Stanley.

Biographical notes

Wollman, Elie

Born in Paris in 1917, the son of Eugène and Elisabeth Wollman, he studied medicine and biological sciences at the University of Paris. Drafted at the beginning of World War II, Wollman joined the French Resistance in 1940. In late 1945 he joined Lwoff's department of microbial physiology at the Institut Pasteur, where he worked on bacteriophage and bacterial genetics, culminating in his collaboration with Jacob on bacterial conjugation and episomes. In the 1960s, Wollman was one of a small group of dedicated members who managed to pull the institute through the severe difficulties in which it had found itself.

Wollman, Elisabeth   1888–1943

Born in Minsk, Mme Wollman studied physical chemistry in Liège, where she married a fellow student from Minsk, Eugène Wollman. In 1910 they began their life together in Paris, where she at first worked with Jacques Duclaux (son of Emile Duclaux) on the application of physical chemistry to biological problems, before beginning her long and fruitful collaboration with her husband in the study of lysogeny.

Wollman, Eugène   1883–1943

Born in Minsk, Wollman in 1902 went to Liège, where he initially studied engineering before committing himself to the study of medicine and biology. In 1910 he joined the laboratory of Metchnikoff in the Institut Pasteur, and after serving in the army medical corps in World War I, he became absorbed in the problems of bacteriophage and, ultimately, lysogeny, which with his wife he came very close to solving before becoming the victim of the senseless destruction of Auschwitz where he was taken with his wife in 1943, and from which they never returned.

(*Annls Inst. Pasteur, Paris*, 1946, **72**: 855–8.)

Youatt, William   1776–1847

An English veterinary surgeon. He was originally educated for the nonconformist ministry, but preferred to devote himself to a veterinary infirmary in London, where he lectured to veterinary students at University College from 1830 to 1835. Youatt founded and edited the journal *Veterinarian*, from 1828. A founder member of the Royal Agricultural Society, he received the diploma of the Royal College of Veterinary Surgeons in 1844. *The dog* was published posthumously in 1850.

Zinke, Georg Gottfried

A doctor of medicine, he practised in Jena in the late eighteenth and early nineteenth centuries, wrote a treatise on rabies published in Jena 1804, which rashly promised a cure for persons bitten by mad dogs, but which also contained the first description of rational transmission experiments carried out with a variety of animal species.

References

Numbers in parentheses at the end of each reference give the page numbers on which that reference is mentioned.

1 Abbé, E. (1873). Beiträge zur Theorie des Mikroskops und der mikroskopischen Wahrnehmung. *Arch. mikrosk. Anat. EntwMech.*, **9**: 413–68. (20)

2 Abercrombie, M. (1961). Ross Granville Harrison 1870–1959. *Biogr. Mem. Fellows R. Soc.*, **7**: 111–26. (69)

3 Ackerknecht, E. H. (1966). Zur Geschichte der Tollwut. *Schweiz. med. Wschr.*, **96**: 746–8. (116; 168)

4 Acqua, C. (1918). Richerche sulla malathia del giallume del baco da seta. *Rc. Ist. bactol. Portici*, **3**: 243. (132)

5 Allard, H. A. (1916). Some properties of the virus of the mosaic disease of tobacco. *J. agric. Res.*, **6**: 649–74. (91; 117; 121)

6 Allard, H. A. (1917). Further studies of the mosaic disease of tobacco. *J. agric. Res.*, **10**: 615–31. (128)

7 Anderson, J. F. and Goldberger, J. (1911). Experimental measles in the monkey: a preliminary note. *Publ. Hlth Rep., Wash.* **26**: 847–8. (149)

8 Anderson, T. F. (1950). Destruction of bacterial viruses by osmotic shock. *J. appl. Physiol.*, **21**: 70. (108)

9 Anderson, T. F. (1966). Electron microscopy of phages. in *Phage and the origins of molecular biology*, J. Cairns, G. S. Stent and J. D. Watson (eds.). N.Y.: Cold Spring Harbor Lab. Quant. Biol. (107; 108)

10 Andrewes, C. H. (1929). Virus III in tissue cultures I. – The appearance of intranuclear inclusions *in vitro*. *Br. J. exp. Path.*, **10**: 188–90. (74)

11 Andrewes, Sir Christopher (1973). *In pursuit of the common cold*. London: William Heinemann Medical Books Ltd. (140; 142)

12 Andrewes, C. H. and Carmichael, E. A. (1930). A note on the presence of antibodies to herpes virus in post-encephalitic and other human sera. *Lancet*, **i**: 857–8. (153)

13 Andrewes, C. H., Laidlaw, P. P. and Smith, W. (1934). The susceptibility of mice to the viruses of human and swine influenza. *Lancet*, **ii**: 859–62. (138)

14 Andrewes, Sir Christopher and Pereira, H. G. (1972). *Viruses of vertebrates*, 3rd edn. London: Baillière, Tindall and Cassell. (137; 145; 146)

15 Andriewsky, P. (1914). L'ultrafiltration et les microbes invisibles I. Communication: la peste des poules. *Zentbl. Bakt. ParasitKde, Abt. I, Orig.*, **75**: 90–3. (43; 94; 118)

References and author index

16 Apostolov, K., Spasic, P. and Bojanic, N. (1975). Evidence of a viral aetiology in endemic (Balkan) nephropathy. *Lancet*, **ii**: 1271–3. (176)

17 Aristotle. *Historia animalium*, vol. IV, book VIII, 604a. (2)

18 Armstrong, J. A. (1968). Relation of PLT organisms to the rickettsiae, with special reference to structure and multiplication. *Acta Virol.*, **12**: 15–17. (145)

19 Aurelianus, Caelius (Soranus of Ephesus). *On acute diseases and on chronic diseases*, ed. and transl. by I. E. Drabkin, 1950. Chicago Univ. Press. (2)

20 Avery, O. T., MacLeod, C. M. and McCarty, M. (1944). Studies on the chemical nature of the substance inducing transformation of pneumococcal types. Induction of transformation by a desoxyribonucleic acid fraction isolated from Pneumococcus Type III. *J. exp. Med.*, **79**: 137–57. (108)

21 Bail, O. (1925). Der Kolistamm 88 von Gildemeister und Herzberg. *Med. Klin.*, **21**: 1271–3. (92)

22 Bailey, L. (1973). Viruses and hymenoptera. In *Viruses and invertebrates*, A. J. Gibbs (ed.). Amsterdam and London. North-Holland Publ. Co. (134)

23 Baker, George (1766). *An inquiry into the merits of a method of inoculating the small-pox, which is now practised in several counties in England*. London: J. Dodsley. (148)

24 Baltimore, David (1970). RNA-dependent DNA polymerase in virions of RNA tumour viruses. *Nature, Lond.*, **226**: 1209–11. (167)

25 Barksdale, L. W. and Pappenheimer, A. M., Jr (1954). Phage–host relationships in nontoxigenic and toxigenic diphtheria bacilli. *J. Bact.*, **67**: 220–32. (115)

26 Baron, J. (1838). *The life of Edward Jenner*. London: Colburn. (6; 155)

27 Bassi, Agostino (1835). *Del mal del segno calcinaccio o moscardino, malattia che afflige i bachi da seta e sul modo di liberarne le bigattaje anche le piu infestate*. Lodi: Orcesi. (132)

28 Bauer, D. J. (1977). *The specific treatment of virus diseases*. Lancaster: MTP Press Ltd. (52)

29 Bawden, F. C. (1964). *Plant viruses and virus diseases*, 3rd edn. N.Y.: Ronald Press Co. (28; 130)

30 Bawden, F. C., Pirie, N. W., Bernal, J. D. and Fankuchen, I. (1936). Liquid crystalline substances from virus infected plants. *Nature, Lond.*, **138**: 1051–2. (123; 127)

31 Beale, H. P. (1929). Immunologic reactions with the tobacco mosaic virus. *J. exp. Med.*, **49**: 919–36. (122)

32 Bechhold, H. (1907). Kolloidstudien mit der Filtrationsmethode. *Z. phys. Chem.*, **60**: 257–318. (18)

33 Bedson, S. P. (1936). Observations bearing on the antigenic composition of psittacosis virus. *Br. J. exp. Path.*, **17**: 109–21. (145)

34 Bedson, S. P., Maitland, H. B. and Burbury, Y. M. (1927). Further observations on foot and mouth disease. A. Experimental studies of immunity in guinea-pigs to foot and mouth disease. *J. comp. Path. Ther.*, **40**: 5–28. (144)

35 Behring, E. (1890). Untersuchungen über das Zustandekommen der Diphtherie–Immunität bei Thieren. *Dt. med. Wschr.*, **16**: 1145–8 (30; 115)

36 Beijerinck, M. W. (1899*a*) Ueber ein Contagium vivum fluidum als Ursache der Fleckenkrankheit der Tabaksblätter. *Zentbl. Bakt. Parasit-Kde*, Abt. II, **5**: 27–33. (First published in 1898 as Over een contagium vivum fluidum als oorzaak van de vlekziekte der tabaksbladen. *Versl. gewone Vergad. wis-en natuurk. Afd. K. Akad. Wet. Amst.*, **7**: 229–35.) (13; 27; 30; 82; 114; 121)

37 Beijerinck, M. W. (1899*b*) Bemerkung zu dem Aufsatz von Herrn Iwanowsky über die Mosaikkrankheit der Tabakspflanze. *Zentbl. Bakt. ParasitKde*, Abt. I, **5**: 310–11. (30)

38 Beijerinck, M. W. (1922). Pasteur en de ultramicrobiologie. *Chem. Weekbl.*, **19**: 525–7. (95)

39 Belloni, L. (1960) Anatomia plastica: the wax models in Florence. *Ciba-Symposium*, **8**: 129–32. (157)

40 Bergold, G. (1947). Die Isolierung des Polyeder-Virus und die Natur der Polyeder. *Z. Naturf.*, **2b**: 122–43. (132)

41 Bergold, G. (1948). Über die Kapselvirus-Krankheit. *Z. Naturf.*, **3b**: 338–42. (133)

42 Bernal, J. D. and Fankuchen, I. (1941). X-ray and crystallographic studies of plant virus preparations. *J. gen. Physiol.*, **25**: 111–65. (124)

43*a* Bert, Paul (1882). Contributions à l'étude de la rage. *C.r. hebd. Séanc. Acad. Sci., Paris*, **95**: 1253–4. (25)

43*b* Beswick, T. S. L. (1962). The origin and the use of the word herpes. *Med. Hist.*, **6**: 214–32. (151)

44 Bigelow, S. L. and Gemberling, A. (1907). Collodion membranes. *J. Am. chem. Soc.*, **29**: 1576–89. (18)

45 Bittner, J. J. (1936). Some possible effects of nursing on mammary gland tumor incidence in mice. *Science, N.Y.*, **84**: 162–3. (161)

46 Black, L. M. (1950). A plant virus that multiplies in its insect vector. *Nature, Lond.*, **166**: 852–3. (128)

47 Blumberg, B. S., Alter, H. J. and Visnich, S. (1965). A new antigen in leukemia sera. *J. Am. med. Ass.*, **191**: 541–6. (171)

48 Bohr, Niels (1933). Light and Life. *Nature, Lond.*, **131**: 421–3. (103)

49 Bollinger, O. (1873). Über Epithelioma contagiosum beim Haushuhn und die sogenannten Pocken des Geflügels. *Virchows Arch. path. Anat. Physiol.*, **58**: 349–61. (44; 154)

50 Bordet, J. (1931). Croonian Lecture. The theories of the bacteriophage. *Proc. R. Soc. Lond., B*, **107**: 398–417. (91; 93)

51 Bordet, J. (1925). Le problème de l'autolyse microbienne transmissible ou du bactériophage. *Annls Inst. Pasteur, Paris*, **39**: 711–63. (92)

52 Bordet, J. and Ciuca, M. (1920). Exsudats leucocytaires et autolyse microbienne transmissible. *C.r.séanc. Soc. Biol.*, **83**: 1293–6. (91; 92; 95)

53 Borrel, A. (1903). Epithélioses infectieuses et épithéliomas. *Annls Inst. Pasteur, Paris*, **17**: 81–118. (79; 158)

54 Boycott, A. E. (1928). The transition from live to dead: the nature of filterable viruses. *Proc. R. Soc. Med.*, **22**: 55–69. (93; 153; 161)

55 Brenner, S., Jacob, F. and Meselson, M. (1961). An unstable intermediate carrying information from genes to ribosomes for protein synthesis. *Nature, Lond.*, **190**: 576–81. (114)

56 Buist, J. B. (1887). *Vaccinia and variola: a study of their life history*. London: J. and A. Churchill. (Also in 1886 in *Proc. R. Soc. Edinb.*, **13**: 603–20.) (8; 21; 154)

57 Burdon Sanderson, J. (1869). Introductory report on 'The intimate pathology of contagion'. In Appendix to *12th Annual Report of the Medical Committee of the Privy Council*. London: Eyre and Spottiswoode. (8; 9)

58 Burkitt, D. (1962). A children's cancer dependent on climatic factors. *Nature, Lond.*, **194**: 232–4. (164)

59 Burnet, F. M. (1934). The bacteriophages. *Biol. Rev.*, **9**: 332–50. (96)

60 Burnet, F. M. (1936). Influenza virus on the developing egg: I. Changes associated with the development of an egg-passage strain of virus. *Br. J. exp. Path.*, **17**: 282–93. (138)

61 Burnet, F. M. (1942). The affinity of Newcastle disease virus to the influenza virus group. *Aust. J. exp. Biol. med. Sci.*, **20**: 81–8. (139)

62 Burnet, Sir M. (1945). *Virus as organism*. Cambridge, Mass.: Harvard Univ. Press. (81)

63 Burnet, F. M. (1966). Men or molecules?. *Lancet*, **i**: 37–9. (81; 100)

64 Burnet, F. M. and McKie, M. (1929). Observations on a permanently lysogenic strain of *B. enteritidis gaertner*. *Aust. J. exp. Biol. med. Sci.*, **6**: 277–84. (96; 111)

65 Burnet, Sir Macfarlane and White, D. O. (1972). *Natural history of infectious disease*. Cambridge: Cambridge Univ. Press. (33; 136; 140; 144)

66 Burrows, M. T. (1910). A method of furnishing a continuous supply of new medium to a tissue culture *in vitro*. *J. Am. med. Ass.*, **55**: 2057–8. (73)

67 Cairns, J., Stent, G. S. and Watson, J. D. (eds.) (1966). *Phage and the origins of molecular biology*. N.Y.: Cold Spring Harbor Lab. Quant. Biol. (100; 104)

68 Carlson, E. A. (1971). An unacknowledged founding of molecular biology: H. J. Muller's contribution to gene theory 1910–1936. *J. Hist. Biol.*, **4**: 149–70. (94)

69 Carrel, A. (1923). A method for the physiological study of tissues *in vitro*. *J. exp. Med.*, **38**: 407–18. (69; 73)

70 Carrel, A. (1924). Action de l'extrait filtré du sarcome de Rous sur les macrophages du sang. *C.r. Séanc. Soc. Biol.*, **91**: 1069–71. (71)

71 Carrel, A. (1926). Some conditions of the reproduction *in vitro* of the Rous virus. *J. exp. Med.*, **43**: 647–68. (71)

72 Carrel, A. and Rivers, T. M. (1927). La fabrication du vaccin *in vitro*. *C.r. Séanc. Soc. Biol.*, **96**: 848–50. (73)

73 Celsus. *De medicina*, English transl. by W. G. Spencer, 1938. London: William Heinemann. (3)

74 Centanni, E. (1902). Die Vogelpest. *Zentbl. Bakt. ParasitKde*, Abt. I, Orig., **31**: 145–52; 182–201. (28; 38; 40; 139)

75 Centanni, E. and Savonuzzi, E. (1901). La peste aviaria I & II, communicazione fatta all'accademia delle scienze mediche e naturali de Ferrara, 9 marzo e 4 aprile, 1901. *Riforma medica*, **17**(ii): 81; 332. (37)

76 Chamberland, C. (1884). Sur un filtre donnant de l'eau physiologiquement pure. *C.r. hebd. Séanc. Acad. Sci., Paris*, **99**: 247–8. (25)

References and author index

77 Chargaff, E. (1971). Preface to a grammar of biology. A hundred years of nucleic acid research. *Science, N.Y.*, **172**: 637–42. (100)

78 Chauveau, A. (1868). Nature du virus vaccin. Détérmination expérimentale des éléments qui constituent le principe actif de la sérosité vaccinale virulente. *C.r. hebd. Séanc. Acad. Sci., Paris*, **66**: 289–93. (8; 154; 155)

79 Churchill, A. E. and Biggs, P. M. (1967). Agent of Marek's disease in tissue culture. *Nature, Lond.*, **215**: 528–30. (164)

80 Churchill, A. E., Payne, L. N. and Chubb, R. C. (1969). Immunization against Marek's disease using a live attenuated virus. *Nature, Lond.*, **221**: 744–7. (164)

81 Churchill, F. B. (1975). Chabry, Roux, and the experimental method in nineteenth century embryology. In *Foundations of scientific method: the nineteenth century*, R. N. Giere and R. S. Westfall (eds.). Bloomington and London: Indiana Univ. Press. (68)

82 Clark, P. F. (1959). Hideyo Noguchi, 1876–1928. *Bull. Hist. Med.*, **33**: 1–20. (145)

83 Cleghorn, George (1774). *Observations on the epidemical diseases in Minorca from the year 1744 to 1749*, 4th edn. London: T. Cadell. (169)

84 Cockayne, E. A. (1912). Catarrhal jaundice: sporadic and epidemic, and its relation to acute yellow atrophy of the liver. *Q. Jl. Med.*, **6**: 1–29. (169)

85 Cohen, S. S. (1947). The synthesis of bacterial viruses in infected cells. *Cold Spring Harb. Symp. quant. Biol.*, **12**: 35–49. (105)

86 Cohen, S. S. and Anderson, T. F. (1946). Chemical studies on host virus interactions I. The effect of bacteriophage adsorption on the multiplication of its host *Escherichia coli* B. *J. exp. Med.*, **84**: 511–23. (105)

87 Cohen, S. S. and Stanley W. M. (1942). The molecular size and shape of the nucleic acid of tobacco mosaic virus. *J. biol. Chem.*, **144**: 589–98. (105)

88 Coley, N. G. (1973). *From animal chemistry to biochemistry*. Amersham: Hulton Educational Publications. (14; 42)

89 Collier, R. (1974). *The plague of the Spanish lady: the influenza pandemic of 1918–1919*. London: Macmillan. (136)

90 Conn, H. J. (1946). Development of histological staining. *Summit, N.J., Ciba Symp.*, **7**: 270–300: (20)

91 Cooper, P. D. (1961). An improved agar cell-suspension plaque assay for poliovirus: some factors affecting efficiency of plating. *Virology*, **13**: 153–7. (77)

92 Copeman, S. M. (1899). *Vaccination: its natural history and pathology* (Milroy Lectures 1898). London: Macmillan. (39)

93 Creighton, C. (1891–94). *A history of epidemics in Britain*. Cambridge: Cambridge Univ. Press. (148; 154)

94 Creutzfeldt, H. G. (1920) Über die eigenartige herdförmige Erkrankung des Zentralnervensystems. *Z. ges. Neurol. Psychiat.*, **57**: 1–18. (175)

95 Crick, F. H. C. and Watson, J. D. (1956). Structure of small viruses. *Nature, Lond.*, **177**: 473–5. (126)

96 Crowther, J. A. (1926). The action of X-rays on *Colpidium colpoda*. *Proc. R. Soc. Lond.*, B, **100**: 390–404. (101)

97 Cuillé, J. and Chelle, P.-L. (1936). La maladie dite tremblante du mouton est-elle inoculable?. *C.r. hebd. Séanc. Acad. Sci., Paris*, **203**: 1552–4. (174)

98 Darlington, C. D. (1944). Heredity, development and infection. *Nature, Lond.*, **154**: 164–9. (124)

99 Davaine, C. (1864). Nouvelles recherches sur la nature de la maladie charbonneuse connue sous le nom de 'sang de rate'. *C.r. hebd. Séanc. Acad. Sci., Paris*, **59**: 393–6. (7)

100 *de Laval, Gustav 1845–1913: de höga hastigheternas man.* Minneskrift utgiven av AB de Lavals Ångturbin. Stockholm: Tekniska Museet, 1943. (18; 19)

101 Delbrück, M. 'Riddle of life', appendix to 'A physicist's renewed look at biology: twenty years later'. *Science, N.Y.*, **168**: 1312–15. (102)

102 Delbrück, M. and Luria, S. E. (1942). Interference between two bacterial viruses acting upon the same host, and mechanism of virus growth. *Archs Biochem.*, **1**: 111–41. (107)

103 de Noronha, F., Baggs, R., Schäfer, W. and Bolognesi, D. P. (1977). Prevention of oncornavirus-induced sarcomas in cats by treatment with antiviral antibodies. *Nature, Lond.*, **267**: 54–6. (66)

104 Dessauer, F. (1922). Ueber einige Wirkungen von Strahlen. *Z. Phys.*, **12**: 38–47. (101)

105 d'Herelle, F. (1917). Sur un microbe invisible antagoniste des bacilles dysentériques. *C.r. hebd. Séanc. Acad. Sci., Paris*, **165**: 373–5. (88)

106 d'Herelle, F. (1921). *Le bacteriophage.* Paris: Masson. (77; 89; 90)

107 d'Herelle, F. (1922). *The bacteriophage.* Baltimore: Williams and Wilkins Co. (88)

108 d'Herelle, F. (1923). Autolysis and bacteriophagis. *J. St. Med.*, **31**: 461–6. (91)

109 d'Herelle, F. (1949). The bacteriophage. *Science News*, **14**: 44–59. (88)

110 Diener, T. O. (1973). Similarities between the scrapie agent and the agent of the potato spindle tuber disease. *Annls clin. Res.*, **5**: 268–78. (176)

111 Diener, T. O. (1967). Potato spindle-tuber virus; a plant virus with properties of a free nucleic acid. *Science, N.Y.*, **158**: 378–81. (176)

112 Dochez, A. R., Shibley, G. S. and Miles, K. C. (1930). Studies in the common cold. IV. Experimental transmission of the common cold to anthropoid apes and human beings by means of a filtrable agent. *J. exp. Med.*, **52**: 701–16. (142)

113 Doermann, A. H. (1948). Intracellular growth of bacteriophage. *Carnegie Inst. Wash. Yb.*, **47**: 176–82. (107)

114 Doermann, A. H. (1952). The intercellular growth of bacteriophages. I. Liberation of intracellular bacteriophage T4 by premature lysis with another phage or with cyanide. *J. gen. Physiol.*, **35**: 645–56. (107)

115 Doerr, R. (1920). Sitzungsberichte der Gesellschaft der schweizerischen Augenärtzte: Diskussion. *Klin. Mbl. Augenheilk.*, **65**: 104. (152)

116 Doerr, R. and Pick, R. (1915). Untersuchungen über das Virus der Hühnerpest. *Zentbl. Bakt. ParasitKde*, Abt. I, Orig., **76**: 476–94. (137)

117 Doerr, R. and Seidenberg, S. (1933). Untersuchungen über das Virus der Hühnerpest. VII. Zu Virusadsorption *in vitro*. *Z. Hyg. Infekt Krankh.*, **114**: 269–75. (137)

118 Doerr, R. (1938). Die Entwicklung der Virusforschung. In *Handbuch der Virusforschung*, vol. I, R. Doerr and C. Hallauer (eds.). Wien: Verlag von Julius Springer (153)

119 Dooren de Jong, L. E. den (1931). Studien über Bakteriophagie I. Ueber *Bac. megatherium* und den darin anwesenden Bakteriophagen. *Zentbl. Bakt. ParasitKde*, Abt. I, Orig., **120**: 1–15. (98)

120 Dooren de Jong, L. E. den (1940). Beijerinck the man. In *Verzamelde Geschriften van M. W. Beijerinck*, vol. VI., pp. 3–47 of 'Martinus Willem Beijerinck. His life and his works'. Gravenhage: Martinus Nijhoff. (27)

121 Doyle, T. M. (1927). Hitherto unrecorded disease of fowls due to filter-passing virus. *J. comp. Path. Ther.*, **40**: 144–69. (137; 139)

122 Duckworth, D. H. (1976). Who discovered bacteriophage?. *Bact. Rev.*, **40**: 793–802. (88)

123 Duggar, B. M. and Armstrong, J. K. (1923). Indications respecting the nature of the infective particles in the mosaic disease of tobacco. *Ann. Mo. bot. Gdn*, **10**: 191–212. (94; 119)

124 Duggar, B. M. and Karrer, J. L. (1921). The sizes of the infective particles in mosaic disease of tobacco. *Ann. Mo. bot. Gdn*, **8**: 345–56. (43; 118)

125 Dulbecco, R. (1950). Experiments on photoreactivation of bacterio-phages inactivated with ultraviolet radiation. *J. Bact.*, **59**: 329–47. (109)

126 Dulbecco, R. (1952). Production of plaques in monolayer tissue cultures by single particles of an animal virus. *Proc. natn. Acad. Sci. U.S.A.* **38**: 747–52. (77; 109)

127 Dulbecco, R. (1963). Transformation of cells *in vitro* by viruses. *Science, N.Y.*, **142**: 932–6. (166)

128 Dulbecco, R. (1966). Plaque technique and animal virology: In *Phage and the origins of molecular biology*, J. Cairns, G. S. Stent and J. D. Watson (eds.). N.Y.: Cold Spring Harbor Lab. Quant. Biol. (77; 109)

129 Dulbecco, R. (1971). From lysogeny to animal viruses. In *Of microbes and life*, J. Monod and E. Borek (eds.). N.Y. and London: Columbia Univ. Press. (99; 100)

130 Dulbecco, R. and Freeman, G. (1959). Plaque production by the polyoma virus. *Virology*, **8**: 396–7. (166)

131 Eddy, B. E., Borman, G. S., Grubbs, G. E. and Young, R. D. (1962). Identification of the oncogenic substance in Rhesus monkey kidney cell cultures as Simian Virus 40. *Virology*, **17**: 65–75. (162)

132 Editorial (1910). Another martyr to science. *Lancet*, **i**: 1506. (80)

133 Editorial (1968). Coronaviruses. *Nature, Lond.*, **220**: 650. (142)

134 Edward, D. G. ff. and Freundt, E. A. (1956). The classification and nomenclature of organisms of the pleuropneumonia group. *J. gen. Microbiol.*, **14**: 197–207. (34; 145)

135 Edwards, W. S. and Edwards, P. D. (1974). *Alexis Carrel: visionary surgeon*. Springfield, Illinois: Ch. C. Thomas. (69)

136 Ehrlich, Paul (1911). Ueber Salvarsan. *Münch. med. Wschr.*, **58** (ii): 2481–6. (170)

137 Elford, W. J. (1928). Ultrafiltration. *Jl. R. microsc. Soc.*, **48**: 36–45. (98)

138 Elford, W. J. (1933). The principles of ultrafiltration as applied in biological studies. *Proc. R. Soc. Lond., B,* **112**: 384–406. (18)

139 Elford, W. J. (1938). The sizes of viruses and bacteriophages, and methods for their determination. In: *Handbuch der Virusforschung,* vol. I, R. Doerr and C. Hallauer (eds.). Wien: Verlag von Julius Springer. (98; 138; 155)

140 Ellerman, V. and Bang, O. (1908). Experimentelle Leukämie bei Hühnern. *Zentbl. Bakt. ParasitKde,* Abt. I, Orig., **46**: 595–609. (47; 159)

141 Ellis, E. L. (1966). Bacteriophage: one step growth. In: *Phage and the origins of molecular biology,* J. Cairns, G. S. Stent and J. D. Watson (eds.). N.Y.: Cold Spring Harbor Lab. Quant. Biol. (103)

142 Ellis, E. L. and Delbrück (1939). The growth of bacteriophage. *J. gen. Physiol.,* **22**: 365–84. (103)

143 Enders, J. F., Weller, T. H. and Robbins, F. C. (1949). Cultivation of the Lansing strain of poliomyelitis virus in cultures of various human embryonic tissues. *Science, N.Y.,* **109**: 85–7. (63; 76)

144 Enders, J. F. and Peebles, T. C. (1954). Propagation in tissue cultures of cytopathogenic agents from patients with measles. *Proc. Soc. exp. Biol. Med.,* **86**: 277–86. (65; 149)

145 Epstein, M. A., Achong, B. G. and Barr, Y. M. (1964). Virus particles in cultured lymphoblasts from Burkitt's lymphoma. *Lancet,* i: 702–3. (165)

146 Erdman, Rhoda (1916). Attenuation of the living agents of cyanolophia. *Proc. Soc. exp. Biol. Med.,* **13**: 189–93. (71; 72)

147 Errera, L. (1906). Sur la limite de petitesse des organismes. *Recl Inst. bot. 'Léo Errera',* **6**: 73–82. (83; 84)

148 *Commémoration Léo Errera 1958,* Université Libre de Bruxelles, 1958. (84)

149 Evans, A. S. (1976). Causation and disease: the Henle–Koch postulates revisited. *Yale J. Biol. Med.,* **49**: 175–95. (67; 79)

150 Everitt, C. W. F. (1975). *James Clerk Maxwell: physicist and natural philospher.* N.Y.: Charles Scribner's Sons. (83)

151 Fawcett, D. W. (1956). Electron microscope observations on intracellular virus-like particles associated with the cells of the Lucké renal adenocarcinoma. *J. biophys. biochem. Cytol.,* **2**: 725–42. (164)

152 Fenner, F. and Ratcliffe, F. N. (1965). *Myxomatosis.* London and N.Y.: Cambridge Univ. Press. (146)

153 Fick, A. (1855). Ueber Diffusion. *Annln Phys., 4.s.,* **94**: 59–86. (17)

154 Findlay, G. C. M. and MacCallum, F. O. (1937). Note on hepatitis and yellow fever immunization. *Trans. R. Soc. trop. Med. Hyg.,* **31**: 297–308. (170)

155 Flaum, A., Malmros, H. and Persson, E. (1926). Eine nosocomiale Ikterus-Epidemie. *Acta med. Scand. Suppls.,* **16**: 544–53. (170)

156 Fleming, D. (1969) Emigré physicists and the biological revolution. In *The intellectual migration: Europe and America 1930–1960,* D. Fleming and B. Bailyn (eds.). Cambridge, Mass.: Harvard Univ. Press. (102)

157 Fleming, D. and Bailyn, B. (eds.) (1969). *The intellectual migration: Europe and America 1930–1960.* Cambridge, Mass.: Harvard Univ. Press. (102)

158 Flexner, S. and Lewis, P. A. (1909). The nature of the virus of epidemic poliomyelitis. *J. Am. med. Ass.,* **53**: 2095. (51)

159 Fracastoro, Girolamo (1546). *De contagione et contagionis morbis et eorum curatione* Transl. and notes by Willmer Cave Wright, 1930. N.Y. and London: Putnam. (2; 5; 11)

160 Fraenkel-Conrat, H. (1956). The role of the nucleic acid in the reconstitution of active tobacco mosaic virus. *J. Am. chem. Soc.*, **78**: 882–3. (125)

161 Franklin, R. E. (1955). Structure of tobacco mosaic virus. *Nature, Lond.*, **175**: 379–81. (126)

162 Freeman, V. J. (1951). Studies on the virulence of bacteriophage infected strains of *Corynebacterium diphtheriae*. *J. Bact.*, **61**: 675–88. (115)

163 Friedrich-Freksa, H. (1940). Bei der Chromosomenkonjugation wirksame Kräfte und ihre Bedeutung für die identische Verdopplung von Nucleoproteinen. *Naturwissenschaften*, **28**: 376–9. (125)

164 Friedrich-Freksa, H., Melchers, G. and Schramm, G. (1946). Biologischer, chemischer und serologischer Vergleich zweier Parallelmutanten phytopathogener Viren mit ihren Ausgangsformen. *Biol. Zbl.*, **65**: 187–222. (125)

165 Frosch, P. (1924). Die Erreger der Maul- und Klauenseuche. *Berl. Tierärztl. Wschr.*, **40**: 341–2. (143)

166 Fukushi, T. (1933). Transmission of the virus through the eggs of an insect vector. *Proc. imp. Acad. Japan*, **9**: 457–60. (128)

167 Fuller, Thomas (1730). *Exanthematologia: or, an attempt to give a rational account of eruptive fevers, especially of measles and smallpox*. London: C. Rivington and S. Austen. (6; 148)

168 Gajdusek, D. C. and Gibbs, C. J., Jr (1971). Transmission of two subacute spongiform encephalopathies of man (kuru and Creutzfeldt–Jakob disease) to New World monkeys. *Nature, Lond.*, **230**: 588–91 (175)

169 Gajdusek, D. C. and Zigas, V. (1957). Degenerative disease of the central nervous system in New Guinea: the endemic occurrence of 'Kuru' in the native population. *New Engl. J. Med.*, **257**: 974–8. (174)

170 Galtier, V. (1879a). Etudes sur la rage. *Annls Méd. vét.*, **28**: 627–39. (56)

171 Galtier, V. (1879b) Etudes sur la rage. *C.r. hebd. Séanc. Acad. Sci., Paris*, **89**: 444–6. (56)

172 Galtier, V. (1881). Les injections de virus rabique dans le torrent circulatoire ne provoquent pas l'éclosion de la rage et semblent conférer l'immunité. La rage peut être transmise par l'ingestion de la matière rabique. *C.r. hebd. Séanc. Acad. Sci., Paris*, **93**: 284–5. (56)

173 Gamgee, John (1866). *The cattle plague with official reports of the International Veterinary Congress held in Hamburg, 1863, and in Vienna, 1865*. London: R. Hardwicks. (8)

174 Gatti, Angelo (1764). *Réfléxions sur les prejugés qui s'opposent aux progrès et à la perfection de l'inoculation*. Brussels: Musier fils. (6)

175 Gatti, A. (1767). *Nouvelles réfléxions sur la pratique de l'inoculation*. Brussels: Musier fils. (6)

176 Gey, G. O. (1933). An improved technic for massive tissue culture. *Am. J. Cancer*, **17**: 752–6. (76)

177 Gey, G. O. and Bang, F. B. (1951). Viruses and cells – a study in tissue culture applications. 1. Cells involved – availability and susceptibility. *Trans. N.Y. Acad. Sci.*, **14**: 15–24. (77)

178 Gierer, A. and Schramm, G. (1956). Die Infektiosität der Nucleinsäure aus Tabakmosaikvirus. *Z. Naturf.*, **11b**: 138–42. (125)

179 Gildemeister, E. and Herzberg, K. (1924). Zur Theorie der Bakteriophagen (d'Herelle Lysine). 6. Mitteilung über das d'Herellesche Phänomen. *Zentbl. Bakt. ParasitKde*, Abt. I, Orig., **93**: 402–20. (92)

180 Gratia, A. (1921). Preliminary report on a staphylococcus bacteriophage. *Proc. Soc. exp. Biol. Med.*, **18**: 217–19. (91)

181 Gregg, N. M. (1941). Congenital cataract following German measles in the mother. *Trans. ophthal. Soc. Aust.*, **3**: 35–46. (66; 150)

182 Groman, N. B. (1953). Evidence for the induced nature of the change from nontoxigenicity to toxigenicity in *Corynebacterium diphtheriae* as a result of exposure to specific bacteriophage. *J. Bact.*, **66**: 184–91. (115)

183 Groman, N. B. (1955). Evidence for the active role of bacteriophage in the conversion of nontoxigenic *Corynebacterium diphtheriae* to toxin production. *J. Bact.*, **69**: 9–15. (115)

184 Gross, Ludwik (1951). 'Spontaneous' leukemia developing in C3H mice following inoculation, in infancy, with AK–leukemic extracts, or AK-embryos. *Proc. Soc. exp. Biol. Med.*, **76**: 27–32. (162)

185 Grüter, W. (1920). Experimentelle und klinische Untersuchungen über den sogenannten Herpes corneae. *Ber. Versamm. dt. ophthal. Ges.*, **42**: 162–7. (152)

186 Guarnieri, G. (1892). Richerche sulla patogenesi ed etiologia dell'infezioni vaccinica e vaiolosa. *Archo Sci. med.*, **16**: 403–24. (154)

187 Guérin, C. (1905). Contrôle de la valeur des vaccins Jenneriens par la numération des éléments virulents. *Annls Inst. Pasteur, Paris*, **19**: 317–20. (76)

188 Gye, W. E. (1925). The aetiology of malignant new growths. *Lancet*, **ii**: 109–17. (159; 161)

189 Gye, William Ewart, obit. (by C. H. Andrewes). *Obit. Not. Fell. R. Soc. Lond.*, 1952–53, **8**: 419–30. (159)

190 Gye, W. E. and Purdy, W. J. (1931). *The cause of cancer.* London: Cassell. (159)

191 Hadlow, W. J. (1959). Scrapie and kuru. *Lancet*, **ii**: 289–90. (174)

192 Hahon, N. (1964). Introductory note to 'Transmission of the virus through the eggs of an insect vector'. In *Selected papers on virology*, N. Hahon (ed.). Englewood Cliffs, N.J.: Prentice-Hall, Inc. (See also reference 166.) (127)

193 Halberstaedter, L. (1915). V. Prowazek. *Dt. med. Wschr.*, **41**: 407–8. (80)

194 Hall, C. E. (1955). Electron densitometry of stained virus particles. *J. biophys. biochem. Cytol.*, **1**: 1–12. (22)

195 Hallauer, C. (1931). Über das Verhalten von Hühnerpestvirus in der Gewebekultur. *Z. Hyg. InfektKrankh.* **113**: 61–74. (74)

196 Halsband, R. (1953). New light on Lady Mary Wortley Montagu's contribution to inoculation. *J. Hist. Med.*, **8**: 390–405. (5)

197 Harde, Edna (1915). A propos de la culture du vaccin. *C.r. Séanc. Soc. Biol.*, **78**: 545–6. (73)

198 Harrison, R. G. (1907). Observations on the living developing nerve fiber. *Proc. Soc. exp. Biol. Med.*, **4**: 140–3. (69)

References and author index

199 Harrison, R. G. (1912). The cultivation of tissues in extraneous media as a method of morpho-genetic study. *Anat. Rec.*, **6**: 181–93. (69)

200 Hartig, R. (1882). *Lehrbuch der Baumkrankheiten*. Berlin: Verlag von Julius Springer. (24)

201 Harvey, A. McGehee (1975). Johns Hopkins – the birthplace of tissue culture: The story of Ross G. Harrison, Warren H. Lewis and George O. Gey. *Johns Hopkins med. J.*, **136**: 142–9. (71)

202 Heberden, William (1768). On the chickenpox. *Med. Trans. Coll. Phys. Lond.*, **1**: 427–36. (148)

203 Heintzel, K. G. E. (1900). *Contagiöse Pflanzenkrankheiten ohne Microben, mit besonderer Berücksichtigung der Mosaikkrankheit der Tabaksblätter*. Inaug. Diss., Friedrich Alexander Universität, Erlangen. (27; 117)

204 Heister, Lorenz (1739). *Institutiones chirurgicae*. (Part II, Chapter CVII). Amsterdam: Janssonius-Waesberge. (6)

205 Heitler, W. (1961). Erwin Schrödinger 1887–1961. *Biogr. Mem. Fellows R. Soc.*, **7**: 221–5. (105)

206 Hektoen, Ludvig (1905). Experimental measles. *J. infect. Dis.*, **2**: 238–55. (148)

207 Helmholz, H. (1874). Die theoretische Grenze für die Leistungsfähigkeit der Mikroskope. *Pogg. Ann.* (Jubelband), **136**: 557–84. (20)

208 Henderson, William (1841). Notice of the molluscum contagiosum. *Edinb. med. surg. J.*, **56**: 213–18. (44)

209 Henle, G., Henle, W. and Diehl, V. (1968). Relation of Burkitt's tumor-associated herpes-type virus to infectious mononucleosis. *Proc. Natn. Acad. Sci. U.S.A.*, **59**: 94–101. (165)

210 Herriot, R. M. (1951). Nucleic acid-free T2 virus 'ghosts' with specific biological action. *J. Bact.*, **61**: 752–4. (108)

211 Hershey, A. D. (1969). Idiosyncrasies of DNA structure. In *Nobel Lectures in Physiology or Medicine 1963–1970*, pp. 417–24. Amsterdam, N.Y. and London: Elsevier. (114)

212 Hershey, A. D. and Chase, M. (1952). Independent functions of viral protein and nucleic acid in growth of bacteriophage. *J. gen. Physiol.*, **36**(i): 39–56. (108; 113; 125)

213 Hertwig, K. H. (1828). Beiträge zu nähern Kenntniss der Wuhtkrankheit oder Tollheit der Hunde. *Hufeland's Jour. pract. Arzneyk. Wundarzneyk.*, **67**: 3–173. (56)

214 Hippocrates. *The medical works of Hippocrates*. Transl. J. Chadwick and W. N. Mann (eds.), 1950. Oxford: Blackwell. (169)

215 Hirst, G. K. (1941). The agglutination of red cells by allantoic fluid of chick embryos infected with influenza virus. *Science, N.Y.*, **94**: 22–3. (137)

216 Hoggan, I. A. (1934). Transmissibility by aphids of the tobacco mosaic virus from different hosts. *J. agric. Res.*, **49**: 1135–42. (128)

217 Holmes, F. O. (1929). Local lesions in tobacco mosaic. *Bot. Gaz.*, **87**: 39–55. (122)

218 Horsfall, F. L. (1965). Some facts and fancies about cancer. *Perspect. Biol. Med.*, **8**: 167–79. (158)

219 Horstmann, D. M. and Beeson, P.B (1975). John Rodman Paul 1893–1971. *Biogr. Mem. natn. Acad. Sci. U.S.A.*, **47**: 323–68. (61)

220 Hoyle, L. (1952). Structure of the influenza virus: the relationship between biological activity and chemical structure of virus fractions. *J. Hyg.*, **50**: 229–45. (138)

221 Hoyle, L. and Fairbrother, R. W. (1937). Antigenic structure of influenza virus; the preparation of elementary body suspensions and the nature of the complement fixing antigen. *J. Hyg.*, **37**: 512–19. (137)

222 Hunger, F. W. T. (1905). Untersuchungen und Betrachtungen über die Mosaik-Krankheit der Tabakspflanze. *Z. PflKrankh.*, **15**: 257–311. (117)

223 Hunger, F. W. T. (1905b). Neue Theorie zur Ätiologie der Mosaikkrankheit des Tabaks. *Ber. dt. bot. Ges.*, **23**: 415–18. (27; 91; 117)

224 Hunter, John (1793). Observations and heads of inquiry on canine madness. *Trans. Soc. improvem. med. chir. knowl.*, **1**: 294–329. (54)

225 Isaacs, A. and Lindenmann, J. (1957). Virus interference. I. The interferon. *Proc. R. Soc. Lond., B*, **147**: 258–67. (66)

226 Ito, Y. (1960). A tumor-producing factor extracted by phenol from papillomatous tissue (Shope) of cottontail rabbits. *Virology*, **12**: 596–601. (163)

227 Ivanovski, D. I. (1892). On two diseases of tobacco. *Sel'. Khoz. Lêsov.*, **169**(2): 108–21; an English translation by J. M. Irvine of the part of this paper which deals with tobacco mosaic disease appears as an appendix to Hughes, S. S. (1972). The origins and development of the concept of the virus in the late nineteenth century. Ph.D. thesis, London University. (13; 25; 115)

228 Ivanovski, D. I. (1899). Ueber die Mosaikkrankheit der Tabakspflanze. *Zentbl. Bakt. ParasitKde*, Abt. II, **5**: 250–4. (30)

229 Ivanovski, D. I. (1903). Über die Mosaikkrankheit der Tabakspflanze. *Z. PflKrankh.*, **13**: 1–41. (30; 31)

230 Jacob, F. (1971). La belle époque. In *Of microbes and life*, J. Monod and E. Borek (eds.). N.Y. and London: Columbia Univ. Press. (111)

231 Jacob, F. and Monod, J. (1961). Genetic regulatory mechanisms in the synthesis of protein. *J. molec. Biol.*, **3**: 318–56. (113; 114)

232 Jakob, A. (1921). Über eigenartigen Erkrankungen des Zentralnervensystems mit bemerkenswertem anatomischen Befunde. *Z. ges. Neurol. Psychiat.*, **64**: 147–228. (175)

233 Jenner, Edward (1798). *An inquiry into the causes and effects of the variolae vaccinae*. London: Sampson Low. (53)

234 Joest, E. (1902). Unbekannte Infektionsstoffe. *Zentbl. Bakt. ParasitKde*, Abt. I, Orig., **31**: 361–84; 410–22. (28)

235 Johnson, James (1942). Biographical sketch of Adolf Mayer, introduction to 'Concerning the mosaic disease of tobacco'. *Phytopath. Class.* **7**: 11–24. (24)

236 Johnson, James, 1886–1952, obit. *Phytopathology*, 1954, **44**: 335–6. (121)

237 Joklik, W. K. (1962). Some properties of poxvirus deoxyribonucleic acid. *J. molec. Biol.*, **5**: 265–74. (155)

238 Jones, A. W., McCusick, V. A., Harper, P. S. and Wuu, K-D. (1971). George Otto Gey (1899–1970): the HeLa cell and a reappraisal of its origin. *Obstet. Gynec., N.Y.*, **38**: 945–9. (76)

239 Jouan, C. and Staub, A. (1920). Etude sur la peste aviaire. *Annls Inst. Pasteur, Paris*, **34**: 343–57. (74; 139)

240 Kausche, G. A., Pfankuch, E. and Ruska, H. (1939). Die Sichtbarmachung von pflanzlichem Virus im Übermikroskop. *Naturwissenschaften*, **27**: 292–9. (106; 127)

241 Keber, F. (1868). Ueber die mikroskopischen Bestandtheile der Pocken-Lymphe. *Virchows Arch. path. Anat. Physiol.*, **42**: 112–28. (8; 22; 154)

242 Klebs, Edwin (1883). Ueber Diphtherie. *Verhandlungen des Congresses für Innere Medicin*. Wiesbaden: Verlag J. F. Bergmann; reprinted in 1940 in *Bull. Hist. Med.*, **8**: 509–22. (115)

243 Kling, C., Petterson, A. and Wernstedt, W. (1912). Experimental and pathological investigation. In *Investigations on epidemic infantile paralysis*, report from the State Medical Institute of Sweden to the XVth International Congress on Hygiene and Demography, Washington. (62)

244 Klug, A. (1974). Rosalind Franklin and the double helix. *Nature, Lond.*, **248**: 787–8. (125)

245 Klug, A. and Caspar, D. L. D. (1960). The structure of small viruses. *Adv. Virus Res.*, **7**: 225–325. (126)

246 Koch, Robert (1876). Die Aetiologie der Milzbrand-Krankheit, begründet auf die Entwicklungsgeschichte der *Bacillus anthracis*. *Beitr. Biol. Pfl.*, **2**: 277–310. (7)

247 Koch, R. (1877). Verfahren zur Untersuchung, zum Conserviren und Photographiren der Bacterien. *Beitr. Biol. Pfl.*, Bd. II, Heft 3: 399–434. (20)

248 Koen, J. S. (1919). A practical method for field diagnosis of swine diseases. *Am. J. vet. Med.*, **14**: 468–70. (135; 136)

249 Köhler, A. (1904). Mikrophotographische Untersuchungen mit ultraviolettem Licht. *Z. wiss. Mikrosk.*, **21**: 129–65; 273–305. (20)

250 Koning, C. J. (1899). Die Flecken-oder Mosaikkrankheit des holländischen Tabaks. *Z. PflKrankh.*, **9**: 65–80. (27)

251 Koprowski, H., Jervis, G. A. and Norton, T. W. (1952). Immune responses in human volunteers upon oral administration of a rodent–adapted strain of poliomyelitis virus. *Am. J. Hyg.*, **55**: 108–26. (64)

252 Krueger, A. P. (1937). The mechanism of bacteriophage production. *Science, N.Y.*, **86**: 379–80. (107)

253 Kunkel, L. O. (1926). Studies on aster yellows. *Am. J. Bot.*, **13**: 646–705. (127)

254 Kunkel, Louis Otto, 1884–1960, obit. (by F. O. Holmes). *Phytopathology*, 1960, **50**: 777–8. (122)

255 Landsteiner, K. and Berliner, M. (1912). Ueber die Kultivierung des Virus der Hühnerpest. *Zentbl. Bakt. ParasitKde*, Abt. I, Orig., **67**: 165–8. (42; 71)

256 Landsteiner, K. and Levaditi, C. (1909*a*). Sur la paralysie infantile expérimentale. *C.r. Séanc. Soc. Biol.*, **67**: 787–9. (51; 72)

257 Landsteiner, K. and Levaditi, C. (1909*b*). La transmission de la paralysie infantile aux singes. *C.r. Séanc. Soc. Biol.*, **67**: 592–4. (51)

258 Landsteiner, K. and Popper, E. (1908). Mikroscopische Präparate von einem menschlichen und zwei Affenrückenmarken. *Wien. klin. Wschr.*, **21**: 1830. (50)

259 Landsteiner, K. and Popper, E. (1909). I. Übertragung der Poliomyelitis acuta auf Affen', *Z. ImmunForsch. exp. Ther.* (Orig.), **2**: 377–90. (50)

260 Landsteiner, K. (and V. K. Russ) (1906). Beobachtungen über das Virus der Hühnerpest. *Zentbl. Bakt. ParasitKde.* Ref., **38**: 540–2; also published in 1906 in *Arch. Hyg.*, **59**: 286–312. (30; 137)

261 Lechevalier, Hubert (1972). Dmitri Iosifovich Ivanovski 1864–1920. *Bact. Rev.*, **36**: 135–45. (30)

262 Lederberg, E. M. (1951). Lysogenicity in *E. coli* K-12. *Genetics*, **36**: 560. (112)

263 Lederberg, E. M. and Lederberg, J. (1953). Genetic studies of lysogenicity in *Escherichia coli*. *Genetics*, **38**: 51–64. (111)

264 Lederberg, J. and Tatum, E. L. (1946). Novel genotypes in mixed cultures of biochemical mutants of bacteria. *Cold Spring Harb. Symp. quant. Biol.*, **11**: 113–14. (112)

265 Le Guyon, Robert (1967). Borrel et la théorie virusale des cancers. *Bull. Acad. natn. Méd.*, **151**: 585–93. (158)

266 Lépine, P. (1949). Felix d'Herelle (1873–1949). *Annls Inst. Pasteur, Paris,* **76**: 457–60. (88)

267 Levaditi, C. (1913*a*). Virus de la poliomyélite et culture de cellules *in vitro*. *C.r. Séanc. Soc. Biol.*, **75**: 202–5. (72)

268 Levaditi, C. (1913*b*). Virus rabique et culture des cellules *in vitro*. *C.r. Séanc. Soc. Biol.*, **75**: 505. (72)

269 Levaditi, C., Nicolau, S. and Schoen, R. (1926). Recherches sur la rage. *Annls Inst. Pasteur, Paris,* **40**: 973–1068. (81; 152)

270 Lewontin, R. C. (1968). Phage and the origins of molecular biology (essay review). *J. Hist. Biol.*, **1**: 155–61. (103)

271 Lipschütz, B. (1909). Ueber mikroskopisch sichtbare, filtrierbare Virusarten (Strongyloplasmen). *Zentbl. Bakt. ParasitKde,* Abt. I, Orig., **48**: 77–90. (80)

272 Lipschütz, B. (1921). Untersuchungen über die Ätiologie der Krankheiten der Herpesgruppe (Herpes zoster, Herpes genitalis, Herpes febrilis). *Archs Derm. Syph.*, **136**: 428–82. (152)

273 Lisbonne, M. and Carrère, L. (1923). Obtention du principe lytique pour le Bacille de Shiga par culture filtrée de *B. coli. C.r. Séanc. Soc. Biol.*, **88**: 724–6. (92)

274 Locke, F. S. (1901). Die Wirkung der Metalle des Blutplasmas und verschiedener Zucker auf das isolirte Säugethierherz. *Zentbl. Physiol.*, **14**: 670–2. (73)

275 Lode, A. and Gruber, F. (1901). Bakteriologische Studien über die Aetiologie einer epidemischen Erkrankung der Hühner in Tirol. *Zentbl. Bakt. ParasitKde,* Abt. I, **30**: 593–604. (41)

276 Lode, A. (1902). Notizen zur Biologie des Erregers der Kyanolophie der Hühner. *Zentbl. Bakt. ParasitKde,* Abt. I, Orig., **31**: 447–51. (41)

277 Loeffler, F. (1884). Untersuchungen über die Bedeutung der Mikroorganismen für die Entstehung der Diphtherie beim Menschen, bei der Taube und beim Kalbe (Dec. 1883). *Mitt. Ksl. Gesdh. amt.* **2**: 421–99. (115)

278 Loeffler, [F.] and Frosch, [P.] (1897). Summarischer Bericht über die Ergebnisse der Untersuchungen der Kommission zur Erforschung der Maul – und Klauenseuche bei dem Institut für Infektionskrankheiten in Berlin. *Zentbl. Bakt. ParasitKde*, Abt. I, **22**: 257–9. (10; 30)

279 Loeffler, F. and Frosch, P. (1898). Berichte der Kommission zur Erforschung der Maul- und Klauenseuche bei dem Institut für Infektionskrankheiten in Berlin. *Zentbl. Bakt. ParasitKde*, Abt. I, **23**: 371–91. (30)

280 Löwenstein, A. (1919). Aetiologische Untersuchungen über den fieberhaften Herpes. *Münch. med. Wschr.*, **66**: 769–70. (152)

281 Lubran, M. (1965). The arrival of 'the disposable'. *Med. News*, 28 May, p. VII. (171)

282a Lucké, B. (1934). A neoplastic disease of the kidney of the frog, *Rana pipiens*. *Am. J. Cancer*, **20**: 352–79. (164)

282b Lucké, B. (1938). Carcinoma in the leopard frog: its probable causation by a virus. *J. exp. Med.*, **68**: 457–68. (164)

283 Luria, S. E. (1950). Bacteriophage: an essay on virus reproduction. *Science, N.Y.*, **111**: 507–11. (108)

284 Luria, S. E. (1966). Mutations of bacteria and of bacteriophage. In *Phage and the origins of molecular biology*, J. Cairns, G. S. Stent and J. D. Watson (eds.). N.Y.: Cold Spring Harbor Lab. Quant. Biol. (104)

285 Luria, S. E. and Anderson, T. F., (1942). The identification and characterization of bacteriophages with the electron microscope. *Proc. natn Acad. Sci. U.S.A.* **28**: 127–30. (106)

286 Luria, S. E. and Delbrück, M. (1942). Interference between inactivated bacterial virus and active virus of the same strain and of a different strain. *Archs. Biochem.*, **1**: 207–18. (107)

287 Luria, S. E., Delbrück, M. and Anderson, T. F. (1943). Electron microscope studies of bacterial viruses. *J. Bact.*, **46**: 57–77. (106)

288 Luria, S. E. and Exner, F. M. (1941). The inactivation of bacteriophages by X-rays; influence of the medium. *Proc. natn. Acad. Sci. U.S.A.* **27**: 370–5. (104)

289 Lürman, A. (1885). Eine Icterusepidemie. *Berl. klin. Wschr.*, **22**: 20–3. (36; 170)

290 Lush, Dora (1940–43). The chick red cell agglutination test with the viruses of Newcastle disease and fowl plague. *J. comp. Path. Ther.*, **53**: 157–60. (139)

291 Lwoff, A. (1953). Lysogeny. *Bact. Rev.*, **17**: 269–337. (93; 99; 100)

292 Lwoff, A. (1957). The concept of virus. *J. gen. Microbiol.*, **17**: 239–53. (100)

293 Lwoff, A. (1966). The prophage and I. in *Phage and the origins of molecular biology*, J. Cairns, G. S. Stent and J. D. Watson (eds.). N.Y.: Cold Spring Harbor Lab. Quant. Biol. (99; 100)

294 Lwoff, A. and Gutmann, A. (1950). Recherches sur un *Bacillus megatherium*. *Annls Inst. Pasteur, Paris*, **78**: 711–39. (111)

295 Lwoff, A., Siminovitch, L. and Kjeldgaard, N. (1950). Induction de la production de bacteriophages chez une bacterie lysogene. *Annls Inst. Pasteur, Paris*, **79**: 815–58. (111)

296 MacCallum, F. O. (1972). Early studies of viral hepatitis. *Br. med. Bull.*, **28**: 105–8. (169)

297 MacCallum, W. G. and Oppenheimer, E. H. (1922). A method for the study of filterable viruses, as applied to vaccinia. *J. Am. med. Ass.*, **78**: 410–11. (20)

298 McClelland, L. and Hare, R. (1941). The adsorption of influenza virus by red cells and a new *in vitro* method for measuring antibodies for influenza virus. *Can. publ. Hlth J.*, **30**: 530–8. (137)

299 M'Fadyean, J. (1900). African horse-sickness. *J. comp. Path. Ther.*, **13**: 1–20. (35)

300 M'Fadyean, J. (1908). The ultravisible viruses. *J. comp. Path. Ther.*, **21**: 58–68; 168–75; 232–42. (42; 49; 79)

301 McKendrick, J. G. (1901). President's address. *Report Br. Ass.*, Sept., pp. 808–16. (82; 84)

302 McKinney, H. H. (1927). Quantitative and purification methods in virus studies. *J. agric. Res.*, **35**: 13–38. (122)

303 Magendie, F. (1821). Expérience sur la rage. *J. Physiol. exp. path.* **1**: 40–6. (54)

304 Maitland, H. B. and Maitland, M. C. (1928). Cultivation of vaccinia virus without tissue culture. *Lancet*, ii: 596–7. (74; 144)

305a Maitland, Hugh Bethune, 1895–1972, obit. (by A. W. Downie). *J. med. Microbiol.*, 1973, **6**: 253–8 (144); also *Lancet*, 1972, i: 271–2. (144)

305b Maramorosch, Karl. (1952). Direct evidence for the multiplication of aster-yellows virus in its insect vector. *Phytopathology*, **42**: 59–64. (128)

306 Marchoux, E. (1908). Culture *in vitro* du virus de la peste aviaire. *C.r. hebd. Séanc. Acad. Sci., Paris*, **147**: 357–9. (42; 72)

307 Marek, J. (1907). Multiple Nervenentzündung (Polyneuritis) bei Hühnern. *Dt. tierärztl. Wschr.*, **15**: 417–21. (163)

308 Martin, C. J. (1896). A rapid method of separating colloids from crystalloids in solutions containing both. *J. Physiol.*, **20**: 364–71. (17)

309 Marton, L. (1968). *Early history of the electron microscope*. San Francisco: San Francisco Press Inc. (106)

310 Maton, W. G. (1815). Some account of a rash, liable to be mistaken for scarlatina. *Med. Trans. R. Coll. Phys.*, **5**: 149–65. (65; 148)

311 Maxcy, K. F. and Rosenau, M. J. (1965). *Preventive medicine and public health*, 9th edn. N.Y.: Meredith. (65)

312 Maycock, W. d'A. (1972). Hepatitis in transfusion services. *Br. med. Bull.*, **28**: 163–8. (173)

313 Mayer, Adolf (1886). Die Mosaikkrankheit des Tabaks. *Landw. Versuchsstat.*, **32**: 450–67. (24; 25)

314 Medical Officer's special article (1943). Homologous serum jaundice. *Lancet*, i: 691–2. (168)

References and author index

315 Mrowka (1912). Das Virus der Hühnerpest ein Globulin. *Zentbl. Bakt. ParasitKde*, Abt. I, Orig. **67**: 249–68. (42; 94)

316 Muller, H. J. (1922). Variations due to change in the individual gene. *Am. Nat.*, **56**: 32–50. (94; 124)

317 Müller, Wilhelm (1871). Zwei Fälle von Epithelioma cylindro-cellulare der Schilddrüse nebst Bemerkungen zur Theorie der Epitheliombildung. *Jena. Z. Med. Naturw.*, **6**: 456–76. (158)

318 Mulvania, M. (1925). Cultivation of the virus of tobacco mosaic by the method of Olitsky. *Science, N.Y.*, **62**: 37. (120)

319 Mulvania, M. (1926). Studies on the nature of the virus of tobacco mosaic. *Phytopathology*, **16**: 853–71. (121)

320 Murphy, J. B. (1931). Discussion of some properties of the causative agent of a chicken tumour. *Trans. Ass. Am. Physns*, **46**: 182–7. (94)

321 Murphy, J. B. and Rous, Peyton. (1912). The behaviour of chicken sarcoma implanted in the developing embryo. *J. exp. Med.*, **15**: 119–32. (75)

322 Murray, J. A. (1911). Cancerous ancestry and the incidence of cancer in mice. *4th Sci. Repts Imp. Cancer Res. Fund, Lond.*, 104–30. (161)

323 Negri, A. (1903). Beitrag zum Studium der Aetiologie der Tollwuth. *Z. Hyg. InfektKrankh.*, **43**: 507–28. (49)

324 Negri, A. (1905). Sulla filtrazione de virus vaccinio. *Lo Sperimentale* (1909). **59**: 679–80. (49)

325 Negri, A. (1909). Über die Morphologie und den Entwicklungscyklus des Parasiten der Tollwut (*Neuroryctes hydrophobiae* Calkins). *Z. Hyg. Infekt-Krankh.*, **63**: 421–40. (49; 81)

326 Negri, L. and Weber, G. (1954). La 'scabbia norvegese' in una cera del 1851 appartenente alla raccolta dell'Istituto di Patologia de Firenze. *Archo 'de vecchi'*, **20**: 893–11. (157)

327 Nelson, Ray (1922). The occurrence of protozoa in plants affected with mosaic and related diseases. *Techn. Bull. Mich. (St. Coll.) agric. Exp. Stn.*, **58**: 1–30. (119)

328 Nobel Foundation and W. Odelberg (eds.) 1972). *Nobel. The man and his prizes*. N.Y. American Elsevier. (121; 123)

329 Nocard, [E.], Roux, [E.], Borrel, [A.], Salimbeni, [A. T.] and Dujardin-Beaumetz [E.]. (1898). Le microbe de la péripneumonie. *Annls Inst. Pasteur, Paris*, **12**: 240–62. (33)

330 Noguchi, H. (1913). Contribution to the cultivation of the parasite of rabies. *J. exp. Med.*, **18**: 314–16. (72)

331 Nordtmeyer, H. (1891). Über Wasserfiltration durch Filter aus gebrannter Infusorienerde. *Z. Hyg. InfektKrankh.*, **10**: 145–54. (17)

332 Northrop, J. H. (1930). Crystalline pepsin II. General properties and experimental methods. *J. gen. Physiol.*, **13**: 739–80. (123)

333 Northrop, J. (1938). Concentration and purification of bacteriophage. *J. gen. Physiol.*, **21**: 335–66. (107)

334 Olby, R. (1970). The macromolecular concept and the origins of molecular biology. *J. chem. Educ.*, **47**: 168–74. (28)

335 Olby, R. (1974). *The path to the double helix*. London: Macmillan. (100; 101; 103; 118; 166)

336 Olitsky, P. K. (1925). Experiments on the cultivation of the active agent of mosaic disease in tobacco and tomato plants. *J. exp. Med.*, **41**: 129–36. (120)

337 Olitsky, Peter K., 1886–1964, obit. (by C.H.A.). *Lancet*, 1964, **ii**: 423. (120)

338 Oppenheimer, C. (1900). *Die Fermente und ihre Wirkungen.* Leipzig: F. C. W. Vogel. (28)

339 Otto, R. and Munter, H. (1921). Zum d'Herelleschen Phänomen. *Dt. med. Wschr.*, **47**: 1579–81. (92)

340 Padan, E. and Shilo, M. (1973). Cyanophages – viruses attacking blue-green algae. *Bact. Rev.*, **37**: 343–70. (86)

341 Paillot, A. (1924). Sur une nouvelle maladie des chenilles de *Pieris brassicae* L. et sur les maladies du noyau chez les insectes. *C.r. hebd. Séanc. Acad. Sci., Paris*, **179**: 1353–6. (132)

342 Paillot, A. (1926). Sur une nouvelle maladie du noyau ou grasserie des chenilles de *Pieris brassicae* et un nouveau groupe de micro-organismes parasites. *C.r. hebd. Séanc. Acad. Sci., Paris*, **182**: 180–2. (133)

343 Panum, P. L. (1847). Beobachtungen über das Maserncontagium. *Virchows Arch. path. Anat. Physiol.*, **1**: 492–512. (149)

344 Parish, H. J. (1968). *Victory with vaccines.* Edinburgh and London: E. & S. Livingstone Ltd. (65)

345 Parker, F. and Nye, R. N. (1925). Studies on filterable viruses I. Cultivation of vaccine virus. *Am. J. Path.*, **1**: 325–35. (73)

346 Paschen, E. (1919). In 'Discussion'. *Hyg. Rdsch.*, **29**: 313–16. (151)

347 Pasteur, L. (1880). De l'atténuation du virus du choléra des poules. *C.r. hebd. Séanc. Acad. Sci., Paris*, **91**: 673–80. (54)

348 Pasteur, L. (1890). La rage. In *Oeuvres de Pasteur*, vol. 6, pp. 672–88. (10)

349 Pasteur, [L.] (1922–39). *Oeuvres de Pasteur.* Reunies par Pasteur Vallery-Radot, vol. 6. Paris: Masson. (23)

350 Pasteur, [L.], Chamberland [C.] and Roux, [E.] (1881). Le vaccin du charbon. *C.r. hebd. Séanc. Acad. Sci., Paris*, **92**: 666–8. (54)

351 Pasteur, [L.], Chamberland [C.] and Roux, [E.] (1884a). Sur la rage. *Bull. Acad. Méd.*, 2.s., **13**: 661–4. (54)

352 Pasteur, [L.], Chamberland [C.] and Roux, [E.] (1884b). Nouvelle communication sur la rage. *Bull. Acad. Méd.*, 2.s., **13**: 337–44. (49)

353 Pasteur, [L.] and Joubert, [J.] (1877). Etude sur la maladie charbonneuse. *C.r. hebd. Séanc. Acad. Sci., Paris*, **84**: 900–6. (10; 15)

354 Paterson, Robert (1841). Cases and observations on the *molluscum contagiosum* of Bateman, with an account of the minute structure of the tumours. *Edinb. med. surg. J.*, **56**: 279–88. (44; 45)

355 Patrick, A. (1965). A consideration of the nature of the English sweating disease. *Med. Hist.*, **9**: 272–9. (136)

356 Paul, J. R. (1971). *A history of poliomyelitis.* New Haven and London: Yale Univ. Press. (50; 60)

357 Pfankuch, E. and Kausche, G. A. (1940). Isolierung und übermikroskopische Abbildung eines Bakteriophages. *Naturwissenschaften*, **28**: 46. (106)

358 Pfeiffer, R. (1892). Vorläufige Mitteilungen über den Erreger der Influenza. *Dt. med. Wschr.,* **18**: 28. (135)

359 Pirie, N. W. (1973). Frederick Charles Bawden 1908–1972. *Biogr. Mem. Fellows R. Soc.* **119**: 19–63. (123)

360 Prince, A. M. (1968). An antigen detected in the blood during the incubation period of serum hepatitis. *Proc. natn. Acad. Sci. U.S.A.* **60**: 814–21. (171)

361 Pylarino, G. (1715). *Nova et tuta variolas excitandi per transplantationem methodus.* Venice: Gabriel Hertz. (5)

362 Rayleigh (Lord) (1896). On the theory of optical images with special reference to the microscope. *Phil. Mag.,* **42**: 167–95. (20)

363 Reed, W. and Carroll, J. (1899). *Bacillus icteroides* and *Bacillus cholerae suis*; preliminary note. *Med. News,* **74**: 513–14. (47)

364 Reed, W. and Carroll, J. (1902). The etiology of yellow fever. *Am. Med.,* **3**: 301–5. (47)

365 Remlinger, Pierre (1903). Le passage du virus rabique à travers les filtres. *Annls Inst. Pasteur, Paris,* **17**: 834–49. (48)

366 Remlinger, P. (1906). Les microbes filtrants. *Bull. Inst. Pasteur, Paris,* **3**: 337–45; 385–92. (44; 48; 49)

367 Rhazes. A treatise on the small-pox and measles (Transl. by W. A. Greenhill in 1847), in *Med. Class.,* 1939, **4**(i): 22–84. (147)

368 Ringer, S. (1880–82). Regarding the action of hydrate of soda, hydrate of ammonia, and hydrate of potash on the ventricle of the frog's heart. *J. Physiol.,* **3**: 195–202. (73)

369 Ritter, J. (1880). Beitrag zur Frage des Pneumotyphus (Eine Hausepidemie in Uster (Schweiz) betreffend). *Dt. Arch. klin. Med.,* **25**: 53–96. (144)

370 Rivers, T. M. (ed.) (1928). *Filterable viruses.* Baltimore: Williams and Wilkins Co. (17; 50; 72)

371 Rivers, T. M. (1930). Infectious myxomatosis of rabbits. *J. exp. Med.,* **51**: 965–76. (160)

372 Rivers, T. M. (1937). Viruses and Koch's postulates. *J. Bact.,* **33**: 1–12. (79)

373 Rivers, T. M. and Tillett, W. S. (1923). Studies on varicella. The susceptibility of rabbits to the virus of varicella. *J. exp. Med.,* **38**: 673–90. (74)

374 Rivolta, S. (1880). In *L'ornitoitria o la medicina degli uccelli domestici e semidomestici,* S. Rivolta and X. Delprato (eds.). Pisa. (44; 154)

375 Roberts, R. S. (1965). A consideration of the nature of the English sweating sickness. *Med. Hist.,* **9**: 385–9. (136)

376 Roche-Lima, H. da (1916). Zur Aetiologie des Fleckfiebers. *Berl. klin. Wschr.,* **53**(i): 567–9. (80)

377 Röhrer, H. (1960). 50 Jahre Forschung auf dem Riems. *Arch. exp. VetMed.,* **9**: 713–31. (143)

378 Roizman, B. (1971). Herpesviruses, man and cancer. In *Of microbes and life,* J. Monod and E. Borek (eds.). N.Y. and London: Columbia Univ. Press. (176)

379 Rooyen, C. E. van and Scott, G. D. (1948). Smallpox diagnosis with special reference to electron microscopy. *Can. J. publ. Hlth,* **39**: 467–77. (155)

380 Röthlin, Otto Mario (1962). Edwin Klebs 1834–1913; Ein früher Vorkämpfer der Bakteriologie und seine Infahrten. *Zürcher medizingeschichtliche Abhandlungen*, Neue Reihe NR6. Zürich: Juris-Verlag. (10; 14)

381 Rous, Peyton (1911). Transmission of a malignant new growth by means of a cellfree filtrate. *J. Am. med. Ass.*, **56**: 198. (46; 159)

382 Rous, Francis Peyton 1879–1970 (by Sir Christopher Andrewes). *Biogr. Mem. Fellows R. Soc.*, 1971, **17**: 643–62. (159)

383 Rous, P. and Beard, J. W. (1935). The progression to carcinoma of virus-induced rabbit papillomas (Shope). *J. exp. Med.*, **62**: 523–48. (160)

384 Rous, P. and Murphy, J. B. (1911). Tumor implantations in the developing embryo. *J. Am. chem. Ass.*, **56**: 741–2. (74)

385 Roux, E. (1903). Sur les microbes dits 'invisibles'. *Bull. Inst. Pasteur, Paris*, **1**: 7–12; 49–56. (27; 49)

386 Roux, E. and Yersin, A. (1888). Contribution à l'étude de la diphthérie. *Annls Inst. Pasteur, Paris*, **2**: 629–61. (26; 115)

387 Roux, Wilhelm (1894–95). Einleitung. *Arch. EntwMech. Org.*, **1**: 1–42. (68)

388 Ruska, H. (1940). Die Sichtbarmachung der Bakteriophagen Lyse im Übermikroskop. *Naturwissenschaften*, **28**: 45–6. (106)

389 Sabin, A. B. (1959). Present position of immunization against poliomyelitis with live virus vaccines. *Br. med. J.*, **i**: 663–80. (64)

390 Sabin, A. B. and Wright, A. M. (1934). Acute ascending myelitis following a monkey bite, with the isolation of a virus capable of reproducing the disease. *J. exp. Med.*, **59**: 115–36. (153)

391 Sanarelli, G. (1897). L'immunité et la sérotherapie contre la fièvre jaune. *Annls Inst. Pasteur, Paris*, **11**: 433–514. (33)

392 Sanarelli, G. (1898). Das myxomatogene Virus. Beitrag zum Studium der Krankheitserreger ausserhalb des Sichtbaren. *Zentbl. Bakt. ParasitKde*, Abt. I, **23**: 865–73. (19; 33)

393 Sanfelice, F. (1897). Ueber die pathogene Wirkung der Blastomyceten. *Z. Hyg. InfektKrankh.*, **26**: 298–322. (44)

394 Sanfelice, F. (1914). Untersuchungen über das *Epithelioma contagiosum* der Tauben. *Z. Hyg. InfektKrankh.*, **76**: 257–81. (46; 94; 117; 153)

395 Sanfelice, F. (1927–28). Ueber die Natur des Virus der Taubenpocke. *Z. ImmunForsch. exp. Ther.*, **54**: 487–95. (46; 153)

396 Sawyer, W. A., Meyer, K. F., Eaton, M. D., Bauer, J. H., Putnam, P. and Schwentker, F. F. (1944). Jaundice in army personnel in the western region of the United States and its relation to vaccination against yellow fever. *Am. J. Hyg.*, **39**; 337–430; **40**: 55–107. (171)

397 Schadewaldt, H. (1975). Die Erstbeschreibung des *Haemophilus influenzae* durch Richard Pfeiffer. *Dt. med. Wschr.*, **100**: 2405–8. (135)

398 Schäfer, W. (1955). Vergleichende sero-immunologische Untersuchungen über die Viren der Influenza und klassischen Geflügelpest. *Z. Naturf.*, **10b**: 81–91. (38; 139)

399 Schäfer, W. (1963). Structure of some animal viruses and significance of their components. *Bact. Rev.*, **27**: 1–17. (138)

400 Schiller, J. (1973). The genesis and structure of Claude Bernard's experimental method. In *Foundations of scientific method: the nineteenth century*, R. N. Giere and R. S. Westfall (eds.). Bloomington and London: Indiana Univ. Press. (7)

References and author index

401 Schlesinger, Max (1933). Reindarstellung eines Bakteriophagen in mit freiem Auge sichtbaren Mengen. *Biochem. Z.*, **264**: 6–12. (98)

402 Schlesinger, M. (1936). The Feulgen reaction of the bacteriophage substance. *Nature, Lond.* **138**: 508–9. (98)

403 Schlesinger, Max, obit. *Lancet*, 1937, i: 232. (98)

404 Schramm, G. (1943). Über die Spaltung des Tabaksmosaikvirus in niedermolekulare Proteine und die Rückbildung hochmolekularen Proteins aus den Spaltstücken. *Naturwissenschaften*, **31**: 94–6. (124; 125)

405 Schrödinger, Erwin (1944). *What is life? The physical aspect of the living cell.* Cambridge: Cambridge Univ. Press. (105)

406 Schumacher, Wilh. (1860). Ueber Membrandiffusion. *Annln Phys.*, **110**: 337–53. (17)

407 Sédillot, C. (1878). De l'influence des découvertes de M. Pasteur sur le progrès de la chirurgie. *C.r. hebd. Séanc. Acad. Sci., Paris*, **86**: 634–40. (10)

408 Shepherd, R. J., Bruening, G. E. and Wakeman, R. J. (1970). Double-stranded DNA from cauliflower mosaic virus. *Virology*, **41**: 339–47. (127)

409 Shope, R. E. (1932). A filtrable virus causing a tumor-like condition in rabbits and its relationship to virus myxomatosum. *J. exp. Med.*, **56**: 803–22. (160)

410 Shope, R. E. (1933). Infectious papillomatosis of rabbits. *J. exp. Med.*, **58**: 607–24. (160)

411 Shope, R. E. (1958). Influenza: history, epidemiology, and speculation. *Publ. Hlth Rep. (Wash.)*, **73**(2): 165–78. (135)

412 Shope, R. E. (1962). Are animal tumor viruses always virus-like?. *J. gen. Physiol.*, **45**: (Suppl.): 143–54. (163)

413 Shope, R. E. (1966). Evolutionary episodes in the concept of viral oncogenesis. *Perspect. Biol. Med.*, **9**: 258–74. (159)

414 Sigurdsson, B. (1954). Rida, a chronic encephalitis of sheep; with general remarks on infections which develop slowly and some of their special characteristics. *Br. vet. J.*, **110**: 341–54. (174)

415 Simpson, D. I. H. and Zuckerman, A. J. (1977). Marburg and Ebola: viruses in search of a relation. *Nature, Lond.*, **266**: 217–18. (176)

416 Smith, K. M. (1973). Insect viruses. In *Viruses and invertebrates*, A. J. Gibbs (ed.). Amsterdam and London: North-Holland Publ. Co. (133)

417 Smith, K. M. and Wyckoff, R. W. G. (1950). Structure within polyhedra associated with insect virus diseases. *Nature, Lond.*, **166**: 861–2. (133)

418 Smith, Wilson (1935). Cultivation of the virus of influenza. *Br. J. exp. Path.*, **16**: 508–12. (140)

419 Smith, W., Andrewes, C. H. and Laidlaw, P. P. (1933). A virus obtained from influenza patients. *Lancet*, ii: 66–8. (138)

420 Smith, W., Andrewes, C. H. and Laidlaw, P. P. (1935). Influenza: Experiments on the immunization of ferrets and mice. *Br. J. exp. Path.*, **16**: 291–302. (138)

421 Snow, John (1849). On the pathology and mode of communication of cholera. *Lond. med. Gaz.*, **9**: (n.s.): 745–52; 923–9. (14)

422 Snow, John (1855). *On the mode of communication of cholera.* 2nd edn, London J. and A. Churchill. (14)

423 Stamp, J. T. (1962). Scrapie: a transmissible disease of sheep. *Vet. Rec.*, **74**: 357–62. (174)

424 Stanley, W. M. (1935). Isolation of a crystalline protein possessing the properties of tobacco-mosaic virus. *Science, N.Y.*, **81**: 644–54. (102; 123)

425 Stanley, Wendell Meredith, 1904–1971, obit. *Nature, Lond.*, 1971, **233**: 149–50. (122)

426 Stent, Gunther, S. (1966). Introduction: waiting for the paradox. In *Phage and the origins of molecular biology*, J. Cairns, G. S. Stent and J. D. Watson (eds.). N.Y.: Cold Spring Harbor Lab. Quant. Biol. (103)

427 Stewart, S. E., Eddy, B. E., Gochenour, A. M., Borgese, N. G. and Grubbs, G. E. (1957). The induction of neoplasms with a substance released from mouse tumors by tissue culture. *Virology*, **3**: 380–400. (162)

428 Stokes, J. S. (1953). Viral hepatitis. *Am J. med. Sci.*, **225**: 349–57. (173)

429 Stokes, A., Bauer, J. H. and Hudson, N. P. (1928). Experimental transmission of yellow fever to laboratory animals. *Am. J. trop. Med.*, **8**: 103–64. (145)

430 Stokes, J., Ruedemann, R. and Lemon, W. S. (1920). Epidemic infectious jaundice and its relation to the therapy of syphilis. *Archs intern. Med.*, **26**: 521–43. (170)

431 Storey, H. H. (1933). Investigations of the mechanism of the transmission of plant viruses by insect vectors. *Proc. R. Soc. Lond, B*, **113**: 463–85. (128; 129)

432 Storey, H. H. and Bottomley, A. M. (1928). The rosette disease of peanuts (*Arachis hypogaea*). *Ann. appl. Biol.* **15**: 26–45. (128)

433 Streissle, G. and Maramorosch, K. (1963). Reovirus and wound-tumor virus: serological cross reactivity. *Science, N.Y.*, **140**: 996–7. (128)

434 Sumner, J. B. (1926). The isolation and crystallization of the enzyme urease. *J. biol. Chem.*, **69**: 435–41. (121)

435 Svedberg, The. and Fåhraeus, R. (1926). New method for the determination of the molecular weight of the proteins. *J. Am. chem. Soc.*, **48**: 430–8. (19)

436 Takahashi, W. N. and Rawlins, T. E. (1932). Method for determining shape of colloidal particles; application in study of tobacco mosaic virus. *Proc. Soc. exp. Biol. Med.*, **30**: 155–7. (127)

437 Temin, H. M. (1964). Nature of the provirus of Rous sarcoma. *Natn Cancer Inst. Monog.* **17**: 557–70. (166)

438 Temin, H. M. and Mizutani, S. (1970). RNA-dependent DNA polymerase in virions of Rous sarcoma virus. *Nature, Lond.*, **226**: 1211–13. (167)

439 Temin, H. M. and Rubin, H. (1958). Characteristics of an assay for Rous sarcoma cells in tissue culture. *Virology*, **6**: 669–88. (165)

440 Theiler, Arnold (1901). Die südafrikanische Pferdesterbe. *Dt. tierärztl. Wschr.*, **9**: 201–3. (36)

441 Theiler, Sir Arnold (1918). Acute liver-atrophy and parenchymatous hepatitis in horses. *5th & 6th Rep. Dir. vet. Res., Dept, Agric. U.S. Afr.*, April 1918. (36; 170)

442 Theiler, M. (1930). Studies on the action of yellow fever virus in mice. *Ann. trop. Med. Parasit.*, **24**: 249–72. (64)

443 Théodoridès, J. (1974*a*). Utilisation du sel et de l'eau de mer dans le traitement ancien de la rage. *C. r. 99ᵉ Congrès nat. Soc. Savantes*, Besançon, 1974, *sciences*, fasc V: 117–22. (2)

444 Théodoridès, J. (1973). Boissier de Sauvages et la rage. *Hist. Sci. Med.*, **7**: 381–6. (2)

445 Théodoridès, J. (1974*b*). Considérations historiques sur la rage. *Arch. int. Claude Bernard*, **5**: 151–60. (56; 116; 168)

446 Tiegel, E. T. (1871). Die Ursache des Milzbrandes. *KorrespBl. schweizer Ärzte*, **1**: 275–80. (15)

447 Timoféeff-Ressovsky, N. W., Zimmer, K. G. and Delbrück, M. (1935). Über die Natur der Genmutation und der Genstruktur. *Nachr. Ges. Wiss. Göttingen*, VI N.F., **1**: 189–241. (101)

448 Timonis, E. (1714). An account, or history, of the procuring of the smallpox by incision, or inoculation; as it has for some time been practised at Constantinople. *Phil. Trans. R. Soc.*, **29**: 72–82. (5)

449 Trentin, J. J., Yabe, Y. and Taylor, G. (1962). The quest for human cancer viruses. *Science, N.Y.*, **137**: 835–41. (162)

450 Truelove, S. C. and Hogben, L. (1947). A documentary study of jaundice associated with syphilis treatment and blood transfusion. *Br. J. soc. Med.*, **1**: 18–32. (171)

451 Twort, F. W. (1915). An investigation on the nature of ultra-microscopic viruses. *Lancet*, **ii**: 1241–3. (87)

452 Twort, Frederick William, 1877–1950, obit. *Br. med. J.*, 1950, **i**: 778–9. (88)

453 Tyrode, M. V. (1910). The mode of action of some purgative salts. *Archs int. Pharmocodyn.*, **20**: 205–23. (73)

454 Tyzzer, E. E. (1906). The histology of the skin lesions in varicella. *Philipp. J. Sci.*, **1**: 349–72. (150)

455 Vallery-Radot, René (1900). *La vie de Pasteur*. Paris: Librairie Hachette. (53; 58)

456 Veale, H. (1866). History of an epidemic of Rötheln, with observations on its pathology. *Edinb. med. J.*, **12**: 404–14. (65; 148)

457 Velpeau, A. A. L. (1854). *Traité des maladies du sein*. Paris: Masson. (157)

458 Vida, Marco Girolamo (1527). *De bombyce*, lib. II. Rome. (131)

459 Vinson, C. G. (1927). Precipitation of the virus of tobacco mosaic. *Science, N.Y.* **66**: 357–8. (122)

460 Vinson, C. G. and Petre, A. W. (1931). Mosaic disease of tobacco II. Activity of the virus precipitated by lead acetate. *Contr. Boyce Thomson Inst. Pl. Res.*, **3**: 131–45. (122)

461 Von Bokay, J. (1909). Über den aetiologischen Zusammenhang der Varizellen mit gewissen Fällen von Herpes zoster. *Wien. klin. Wschr.*, **22**: 1323–6. (150)

462 Von Economo, C. (1917). Encephalitis lethargica. *Wien. klin. Wschr.*, **30**(i): 581–5. (152)

463 von Magnus, P. (1951). Propagation of the PR' strain of influenza A virus in chick embryos II. The formation of 'incomplete' virus following inoculation of large doses of seed virus. *Acta path. microbiol. scand.*, **28**: 278–93. (137)

464 von Prowazek, S. (1907). Chlamydozoa. *Arch. Protistenk.*, **10**: 336–58 and Chlamydozoa II, **10**: 358–64. (22; 79; 80; 132; 133)

465 von Prowazek, S. (1905). Untersuchungen über die Vaccine. *Arb.K. GesundhAmt.*, **22**: 535–56. (155)

466 Waldmann, O. and Pape, J. (1920). Die künstliche Übertragung der Maul- und Klauenseuche auf das Meerschweinchen. *Berl. tierärztl. Wschr.*, 519–20. (143)

467 Walker, M. N. (1926). A comparative study of the mosaic diseases of cucumbers, tomato and Physalis. *Phytopathology*, **16**: 431–58. (122)

468 [Waterson, A. P.] (1974). Kuru, Creutzfeldt–Jakob, and scrapie. *Lancet*, **ii**: 1551–2. (175)

469 Waterson, A. P. (1976). Infectious particles in hepatitis. *Ann. Rev. Med.*, **27**: 23–35. (173)

470 Watson, J. D. (1966). Growing up in the phage group. In *Phage and the origins of molecular biology*, J. Cairns, G. S. Stent and J. D. Watson (eds.). N.Y.: Cold Spring Harbor Lab. Quant. Biol. (105)

471 Watson, J. D. and Crick, F. H. C. (1953). A structure for deoxyribose nucleic acid. *Nature, Lond.*, **171**: 737–8. (109; 166)

472 Watson, M. A. (1973). Plant viruses. In *Viruses and vertebrates*, A. J. Gibbs (ed.). Amsterdam and London: North-Holland Publ. Co. (130)

473 Weller, T. H. and Neva, F. A. (1962). Propagation in tissue culture of cytopathic agents from patients with rubella-like illness. *Proc. Soc. exp. Biol. Med.*, **111**: 215–25. (65)

474 Wilkinson, L. (1976). The development of the virus concept as reflected in corpora of studies on individual pathogens. 3. Lessons of the plant viruses – tobacco mosaic virus. *Med. Hist.*, **20**: 111–34. (11; 25; 124)

475 Wilkinson, L. (1977). The development of the virus concept as reflected in corpora of studies on individual pathogens. 4. Rabies – two millenia of ideas and conjecture on the aetiology of a virus disease. *Med. Hist.*, **21**: 15–31. (48; 54; 168)

476 Wilkinson, L. and Waterson, A. P. (1975). The development of the virus concept as reflected in corpora of studies on individual pathogens. 2. The agent of fowl plague – a model virus? *Med. Hist.*, **19**: 52–72. (37; 38; 137)

477 Wolff, E. *et al.* (1967). *Philosophie et methodologie scientifiques de Claude Bernard.* Paris: Masson. (7)

478 Wollman, E. (1920). A propos de la note de MM. Bordet et Ciuca. *C.r. Séanc. Soc. Biol.*, **83**: 1478–9. (95)

479 Wollman, E. (1925). Recherches sur la bactériophagie (phénomène de Twort–d'Herelle). *Annls. Inst. Pasteur, Paris*, **39**: 789–832. (96)

480 Wollman, E. (1935). The phenomenon of Twort–d'Herelle and its significance. *Lancet*, **ii**: 1312–14. (87; 96)

481 Wollman, E. and Wollman, E. (1937). Les 'phases' des bactériophages (facteurs lysogènes). *C.r. Séanc. Soc. Biol.*, **124**: 931–4. (98)

482 Wollman, E. and Wollman, E. (1938). Recherches sur le phénomène de Twort–d'Herelle (bactériophagie ou autolyse hérédo-contagieuse). *Annls Inst. Pasteur, Paris*, **60**: 13–57. (98)

483 Wollman, Eugène, and Mme (Elisabeth) Wollman, obit. (by Nicolle). *Annls Inst. Pasteur, Paris*, 1946, **72**: 855–8. (99)

484 Wollman, E. L. (1953). Sur le déterminisme génétique de la lysogénie. *Annls Inst. Pasteur, Paris*, **84**: 281–93. (111)

485 Wollman, Elie L. (1966). Bacterial conjugation. In *Phage and the origins of molecular biology*, J. Cairns, G. S. Stent and J. D. Watson (eds.). N.Y.: Cold Spring Harbor Lab. Quant. Biol. (111)

486 Wollman, E. L. and Jacob, F. (1955). Sur le méchanisme du transfer de matériel génétique au cours de la recombinaison chez *E. coli* K 12. *C.r. hebd. Séanc. Acad. Sci., Paris*, **240**: 2449–51. (113)

487 Woodcock, H. M. (1931). The nature of viruses. *Lancet*, **ii**: 936. (116)

488 Woodruff, A. M. and Goodpasture, E. W. (1931). The susceptibility of the chorioallantoic membrane of chick embryos to infection with the fowl-pox virus. *Am. J. Path.*, **7**: 209–22. (41; 74; 77)

489 Woods, A. F. (1899). The destruction of chlorophyll by oxidizing enzymes. *Zentbl. Bakt. ParasitKde* Abt. II, **5**: 745–54. (91; 116)

490 Wyatt, H. V. (1972). When does information become knowledge? *Nature, Lond.*, **235**: 86–9. (109)

491 Xeros, N. (1954). A second virus disease of the leatherjacket, *Tipula paludosa*. *Nature, Lond.*, **174**: 562–3. (133)

492 Youatt, W. (1851). *The dog*. London: Longmans. (56)

493 Zamecnik, P. C., Keller, E. B., Littlefield, J. W., Hoagland, M. B. and Loftfield, R. B. (1956). Mechanism of incorporation of labelled amino acids into protein. *J. cell. comp. Physiol.*, **47** (Suppl. 1): 81–101. (113)

494 Zinder, N. D. and Lederberg, J. (1952). Genetic exchange in *Salmonella*. *J. Bact.*, **64**: 679–99. (112)

495 Zinke, Georg G. (1804). *Neue Ansichten der Hundswuth, ihrer Ursachen und Folgen, nebst einer sichern Behandlungsart der von tollen Thieren gebissenen Menschen. Für Ärzte und Nichtärzte bestimmt*. Jena: C. E. Gabler. (54; 55)

Subject index